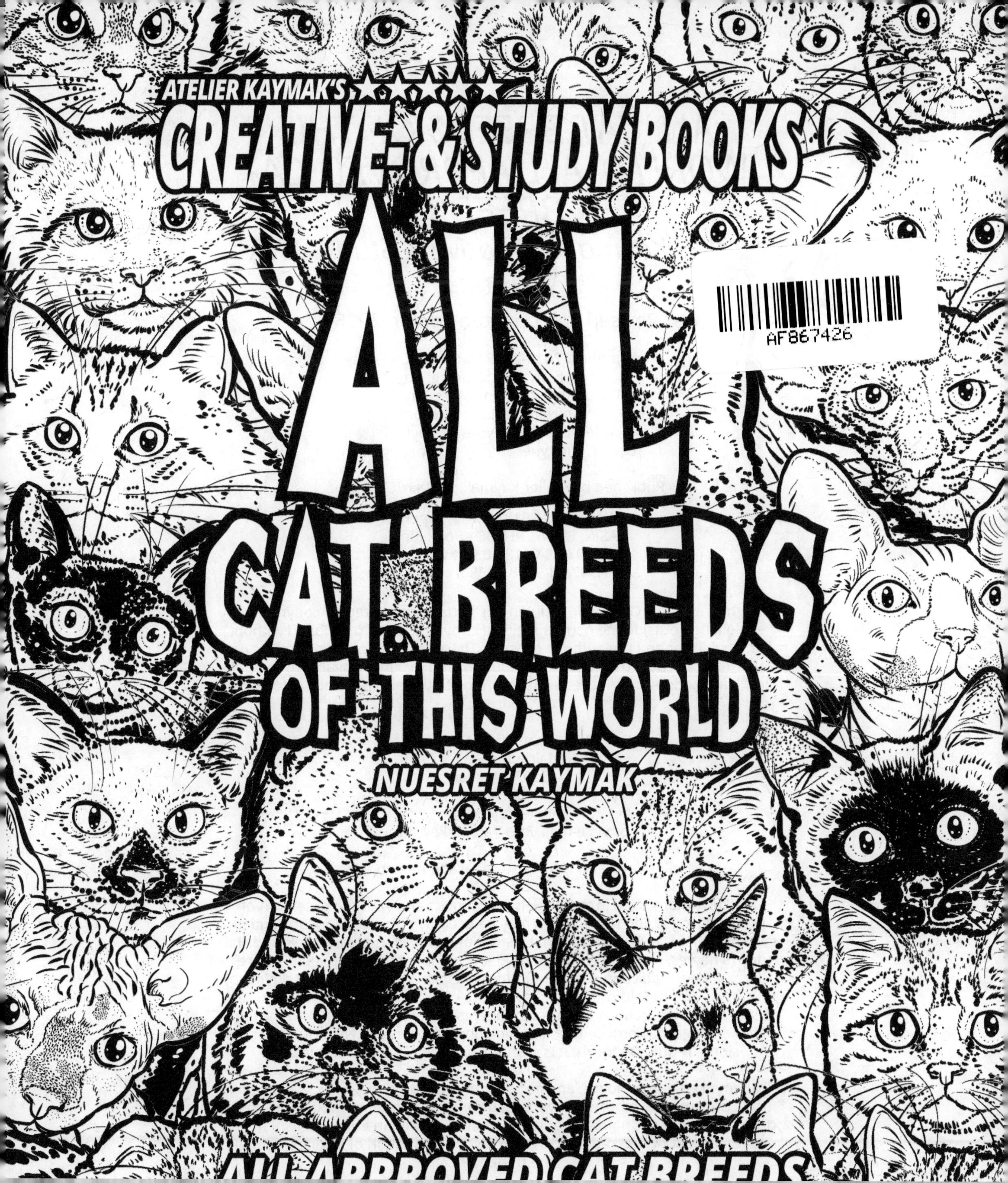
ATELIER KAYMAK'S
CREATIVE- & STUDY BOOKS
ALL
CAT BREEDS
OF THIS WORLD
NUESRET KAYMAK
ALL APPROVED CAT BREEDS

ALL CAT BREEDS OF THIS WORLD - ALL APPROVED CAT BREEDS

Atelier Kaymak's Creative & Study Books Volume 2

Revised and updated edition 2025

Illustration and text: Nuesret Kaymak
Translation & Editing: Alice Kaymak
Proofreading: Ernesto Mora
Title illustration: Nuesret Kaymak
Layout and typesetting: Atelier Kaymak

Published by Atelier Kaymak, Aachen/Germany

ISBN: 978-3-96183-031-2 Paperback
ASIN: B07WLNH19C eBook Kindle
ISBN: 978-3-96183-038-1 eBook PDF
ISBN: 978-3-96183-039-8 eBook ePub

Atelier Kaymak/Nuesret Kaymak
Karl-Friedrich-Str. 176
52072 Aachen
Germany
atelierkaymak.de
or via e-mail: info@atelierkaymak.de

If you discover any errors or have any suggestions or comments about the book, please feel free to write to us at info@atelierkaymak.de.
Dare: every attentive detector gets a nice reward! :-)

ak
ATELIER KAYMAK®

CONTENTS

Preface ... 6
The Fur Categories ... 12
The World's Largest Registries for Pedigree Cats ... 18
Abyssinian ... 24
American Burmese ... 26
American Bobtail Shorthair ... 28
American Bobtail Longhair ... 30
American Curl Shorthair ... 32
American Curl Longhair ... 34
American Shorthair ... 36
American Tiffany ... 38
American Wirehair ... 40
Anatoli ... 42
Aphrodite Shorthair ... 44
Aphrodite Longhair ... 46
Arabian Mau ... 48
Asian also Malayan ... 50
Australian Mist ... 52
Australian Tiffanie ... 54
Balinese ... 56
Bambino ... 58
Bengal ... 60
Burmese cat ... 62
Bohemian Rex ... 64
Bombay ... 66
Brazilian Shorthair ... 68
British Shorthair ... 70
British Longhair ... 72
Burmilla Shorthair ... 74
Burmilla Longhair ... 76
Cashmere ... 78
California Spangled ... 80
Celtic also European Shorthair ... 82
Ceylon Cat alias Sri Lankan Cat ... 84
Chartreux ... 86
Chausie ... 88
Cheetoh ... 90
Classicat also Jungala ... 92
Clipper Cat Shorthair ... 94
Clipper Cat Longhair ... 96
Colorpoint Shorthair ... 98
Colourpoint Longhair alias Himalayan ... 102
Cornish Rex ... 104
Cymric ... 106
Devon Rex ... 108
Domestic Shorthair Cat ... 110
Don Sphynx alias Donskoy ... 112
Dragon Li alias China Li Hua Cat ... 114
Egyptian Mau ... 116
Elf Cat ... 118
European Burmese Cat ... 120
Exotic Shorthair ... 122
Foldex Shorthair ... 124
Foldex Longhair ... 126
Foreign white Shorthair ... 128
Foreign white Longhair ... 130
German Longhair ... 132
German Rex ... 134
Havana also Havana Brown ... 136
Highland Fold ... 138
Highlander Shorthair ... 140
Highlander ... 142
Isle of Man Shorthair ... 144

Isle of Man Longhair ... 146
Japanese Bobtail Shorthair ... 148
Japanese Bobtail Longhair ... 150
Kanaani ... 152
Karelian Bobtail Shorthair ... 154
Karelian Bobtail Longhair ... 156
Khao Manee ... 158
Korat ... 160
Kurilian Bobtail Shorthair ... 162
Kurilian Bobtail Longhair ... 164
Lambkin also Nanus Rex ... 166
LaPerm Shorthair ... 168
LaPerm Longhair ... 170
Lykoi Shorthair Cat alias Werewolf Cat ... 172
Lykoi Longhair Cat alias Werewolf Cat ... 174
Maine Coon also American ... 176
Mandalay ... 180
Manx ... 182
Mekong Bobtail ... 184
Minskin ... 186
Munchkin Shorthair ... 188
Munchkin Longhair ... 190
Nebelung ... 192
Neva Masquarade ... 194
Norwegian Forest Cat ... 196
Ocicat ... 198
Ojos Azules Shorthair ... 200
Ojos Azules Longhair ... 202
Oriental Shorthair ... 204
Oriental Longhair also Javanese ... 206
Persian ... 208
Peterbald ... 210
Pixiebob Shorthair ... 212
Pixiebob Longhair ... 214
Polycoon ... 216
Savannah ... 224
Scottish Fold Shorthair ... 226
Scottish Fold Longhair ... 228
Scottish Shorthair ... 230
Scottish Longhair ... 232
Scottish Straight ... 234
Selkirk Rex Shorthair ... 236
Selkirk Rex Longhair ... 238
Serengeti ... 240
Seychellois Shorthair ... 242
Seychellois Longhair ... 244
Shih Tzu ... 246
Siamese ... 248
Siberian Cat ... 250
Singapura ... 252
Skookum Shorthair ... 254
Snowshoe ... 256
Sokoke ... 258
Somali ... 260
Sphynx ... 262
St Helen Cat Shorthair ... 264
St Helen Cat Longhair ... 266
Suphalak ... 268
Tibetan ... 270
Thai also Wichien Maat ... 272
Tiffanie ... 274
Tonkinese Shorthair ... 276
Tonkinese Longhair ... 278
Toybob ... 280
Toyger ... 282
Turkish Angora ... 284
Turkish Van ... 286
Turkish Vankedisi alias Turkish Van White ... 288
Ukrainian Levkoy ... 290
Ukrainian Levkoy Straight ... 292
Ural Rex Shorthair ... 294

PREFACE

According to scientific knowledge, cats have been around the world for more than 25 million years. Preforms of big cats, saber-toothed tigers, and prehistoric cats developed from prehistoric cat-like carnivores - from which cats of prey, wild cats, and finally domestic cats were formed. In the history of the domestic cat, the time is decisive, in which cats were domesticated for the first time by humans: when humans gave up their so far led life as wandering nomads, as hunters and gatherers - and began to settle down. They became farmers and began to feed on agriculture and animal husbandry. As soon as they started keeping supplies, they were plagued by pests, mice, and rats. The remedy was not long in coming because she was there when needed: cats!

Their pronounced hunting instinct was successfully used to fight rodents. The seemingly insatiable hunger for the unpleasant parasites quickly made the playful cats the chosen heroes and favorites of that time. In this way, the food stores protected attracted starving people, who also settled and came together in the growing settlements. Civilizations developed and shaped the country and its people. With increasing prosperity, important advanced cultures flourished, which made history.

The myths about the cats in ancient Egypt, where the animals were revered and highly valued, are surrounded by legends. More than that, they were even worshipped. Siamese cats in advanced South Asian cultures, Persian cats in the Orient, and Angora cats in Anatolia - united the people in their enjoyment of these lovely animals. The enthusiasm for cats has remained unbroken ever since.

Only with the onset of Christianity did it overturn into the opposite when the Church demonized cats with all the rules of art. Especially black cats were declared incarnations of evil. The intelligent cats preferred to choose animal-loving people and became their roommates. Single women in particular, of course, appreciate such a company. But those women were often a thorn in the side of the bonnet because of their refusal to live alone in the intelligence, or all together, especially among grim, jealous wives in the village. The Inquisition offered remedy and quick disposal. The women were quickly denounced and declared the scapegoat of all possible and impossible. The poor creatures landed, often unsuspectingly by whom they were reported - with their cats - in front of the Grand Inquisitor, who under severe torture forced them to confess that they had always been a witch, to celebrate parties with their cats at night and to wreak havoc everywhere. This was the case for a long time in Europe, especially in Germany and Eastern Europe. This period was probably also called the Dark Middle Ages for this reason.

But modern times, enlightenment and modernity were able to fix the damaged reputation of cats with the victory of reason over prejudice, stupidity, and superstition. The people recognized the valuable helper in the cat with a clear view. They began to appreciate and love cats.

Since then, the triumphal march of cats as man's most popular domestic animal has continued unabated in the western hemisphere. On all continents - except Antarctica - there are cat owners and cat breeders in internationally recognized cat lover associations with regular exhibitions and competitions. They are also the sharp contact points for breeders who have new cat species examined. Each cat is then first put through its paces before it is registered or even declared a new breed. Their most beautiful specimens will then be awarded once a year in a new category. In the course of time, many new species and designations have formed, which are handled in the same or different way by the various cat lover associations in several countries. This can be very confusing for some people and can even be a bit irritating.

This book is intended to provide the cat lover for the first time with an overview of all officially recognized species of the 15 most significant cat lover associations and cat registries in the world.

At the time of research for this book, 136 different cat species were recognized and registered by the World Cat Federation (WCF) - believe it or not. A total of 107 species were listed as official cat breeds. In this revised edition we have extended the prices from the original 31 cats to now all cats. The market price for the respective pedigree cat is based on real sales offers on the platforms gokitty.com (USA), gumtree.com (UK) and quoka.de (D) as catbreedslist.com (CN), mascotarios.org (USA), perfectpets.com.au (AU), thepricer.org (USA) and ukpets.com (UK). The lower price ranges are mostly mixed breeds, kittens or cats that are not attractive for breeders. Purebred cats start at several hundred euros, whereas superstars from breeders can easily cost the price of a luxury car.

So, what can you do with this book?

First of all, get to know all the officially registered cats in the world, which is quite a bit of work. Then you can colour in your cats, cut them out and make them fun for others. You can also cut out and iron the cats onto your T-shirts or bags using iron-on transfer films, which you can buy anywhere. Or you can cut the pages apart and hang them on the wall. Decorate your room with pretty cats... there are many creative ways.

But you must not make big money with the cats. In no case in a commercial form. This does not apply to the institutions mentioned in the imprint: they are also allowed to make money with them on a small scale.

In any case, I wish you a lot of fun and an entertaining time with this creative book.

Sincerely,

Nuesret Kaymak

Aachen, 2023

PS: By the way, we are working on many more interesting titles in the ***SCIENCE PAPERS*** series: ***Evolution***, ***Religion***, ***Dinosaurs***, ***Lions***, ***History of the Cat*** and ***The Family of Big Cats***. Look forward to the new titles!

PPS: Please...PLEASE!!! recommend us to others and support the work of small publishers.

PPPS: A review on the web is quick, costs nothing and helps us immensely! Please be so kind. :-))

THE FUR CATEGORIES

As with all mammals, the skin is one of the most important organs. It is a tissue of cell layers that holds body parts and bones together like a natural vessel. But it is also an effective protective cover that continuously protects the sensitive organs from the lurking dangers of its surrounding external world. For this reason, the skin is one of the most robust and resistant organs of the entire body. Healthy skin can regenerate incredibly quickly and thus repair injuries such as cutting wounds or burns to some extent.

Countless nerve endings in almost all areas of the skin support them. They act as a gigantic network of radar dishes and are in constant contact with the brain. They continuously report on the well-being of the body and the environment in which it currently finds itself. The skin is, therefore, a very effective early warning system, which must constantly defend itself against external influences and intruders such as bacteria and insects.

IIt reacts quickly to temperature fluctuations and signals when things get hot. Nature has come up with something to relieve the skin and give it a remedy: the body hair or the coat. A coat dampens skin and bones from impacts and protects against heat and frost. For a long time, nature, with its selection, has led to the adaptation of animals to their environment, determining their genes.

So, it is understandable that cats in icy surroundings wear correspondingly long fur, and in hot countries, correspondingly short fur. Of course, there are also genetic *accidents* in nature from time to time in the form of mutations, such as albinism and others.

Now man also comes into play as a cat breeder. Through special crosses with wild cats and selected breeds, humans created new cat species, such as the Bengal (a hybrid of a big cat and a domestic cat), the Munchkin (a mutation called the dwarf cat or dachshund cat), the Peke-Face Persian (a mutation called the pig-faced Persian), the Sphynx (the hairless cat) and the Lykoi (a partial hairless cat with a werewolfish appearance).

As exotic and attractive as all these new cats may seem - they give animal protectionists cause for criticism: some cats suffer from permanent pain in organs and joints or breathlessness due to overbreeding. Therefore, some breeds are not allowed in every country: in some countries, such as Germany, animal welfare organizations have enforced a ban on *torture* breeds.

What a cat looks like - determine its genes. The newborn cat carries the genetic material of its parents in the form of genes in the DNA. This information is crucial for further development and shapes the appearance of the cat.

They control the development of the characteristics that form the external and internal characteristics of the cat. Depending on whether the father's or mother's gene is dominant, the cat's appearance develops, especially its coat. Shorthair genes usually dominate over long-hair genes. To which race a cat belongs will determine the length of the fur hair.

The World cat Federation, for example, divides hair length and thus cat breed into four Categories:

- Longhair,
- Medium Longhair (or semi longhair)
- Shorthair and
- Siam-Oriental Shorthair

In this book, we divide the coat length into two Categories for simplicity: Shorthair and longhair.

Not every cat association recognizes the standards of other associations. Therefore, there can be differences here and there in the consideration and classification of individual cats. A famous example is the story of the Ragdoll cat, which occupied the cat associations for decades.

In this book, we have created a small overview that gives information about who accepts which cat and who does not:

ACF	ACFA	CAA	CCA-AFC	CCC	CCCA	CFA	FB	FIFe	GCCF	LOOF	NZCF	SACC	TICA	WCF
✓	✓	✓	✓	✓	✓	✓	✓	✓	✓	✓	✓	✓	✓	✓

Additional orientation aids are the pictograms and symbols, which offer completely uncomplicated additional information about the cats:

Breeding start 1868 United Kingdom	The year and place of the official start of the breeding
BOBTAIL CAT	This cat shows a mutation on the tail due to a gene defect.
Weight 7-16 lbs	The minimum and maximum weight of the cat
Recognized since 1882	The year of the first public recognition
REX CAT	A gene defect causes the cat to grow mutated hair with waves or curls.
Dwarf Cat	This cat has grown very small due to a genetic defect.
ANCIENT ORIGIN	This cat breed can look back on an old history of the origin
HAIRLESS CAT	The cat has no or very few hairs due to a gene defect.
BIG CAT	This cat has a large body.
TOP 10	This cat is one of the ten most popular cat breeds in the world.
HYBRID	The cat is a descendant of a cross between a wild cat and a domestic cat.
LIFE EXPECTATION 9-15 YEARS	The lifespan of the cat
LUXUS	This cat is rare and costs much money because of the high demand.
DESIGNERCAT	This cat is the result of experimental breeding
	The cat has health problems because of hereditary diseases
	This cat may be kept

THE WORLD'S LARGEST REGISTERS FOR PUREBRED CATS

The cat has long outstripped the former champion among pets, the dog. Animal feed manufacturers have calculated that 74.4 million cats were kept in Europe in 2017. Germany leads the table with 13.7 million cats, followed by France with 13.5 million and the UK with 7.9 million. In Russia, 38.4 million cats are kept as pets, 76.5 million in the USA, and 53 million in China. Of course, cat owners everywhere join forces and found more minor associations and clubs - which grow together to form significant associations.

The oldest associations of cat lovers were founded in the 19th century. In the associations, old and new cat breeds are classified, Catalogued, and maintained as standards. Of course, there are numerous associations that are formed in all parts of the world and are certainly justified. We have put together a manageable selection of the most important associations, which should provide a good overview representative of all the many cat associations.

N
W E
S

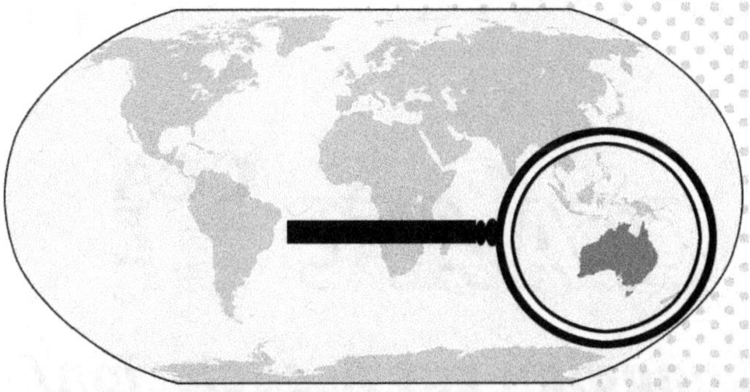

AFC

Australian Cat Federation Inc.

The Australian Association was founded in 1972 in Adelaide and is based in Canning Vale, south of Perth.

Website: https://acf.asn.au

ACFA

American Cat Fanciers Association

The Association of American Cat Lovers was founded in 1955 in the USA and is based in Nixa, Missouri.

Website: https://acfacat.com

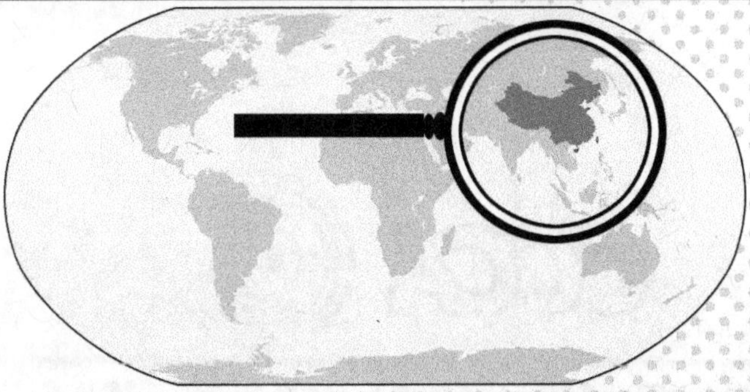

CAA

Cat Aficionado Association

The Chinese Association was founded

CCA-AFC

Canadian cat Association/ Association Féline Canadienne

The association was founded in Canada in 1960 and is based in Mississauga, Ontario.

Website: https://cca-afc.com

CCC

ChatsCanadaCats

Chat Canada cats

The Canadian association was founded in Québec in 2012 and is based in Terrebonne, Montreal.

Website: https://chatscanadacats.ca

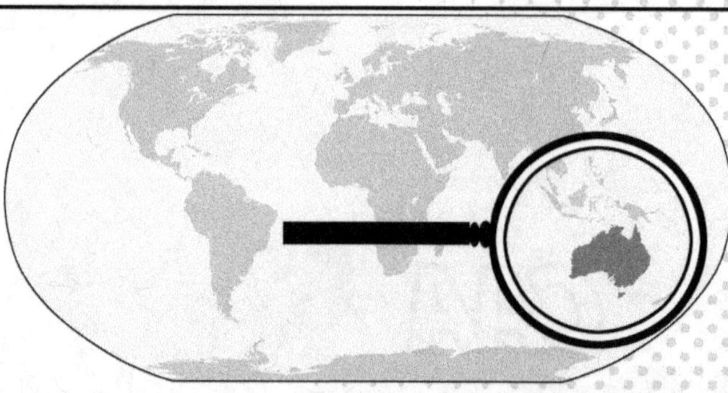

CCCA

CO-ORDINATING CAT COUNCIL OF AUSTRALIA

Co-Ordinating cat council of Australia

The Australian Federation was founded in 1980 in Sydney and is based in Lynham, Canberra.

Website: https://cccofa.com.au

CFA

Cat Fanciers' Association

Founded in Manasquan, New Jersey in 1906, the US association is based in Alliance, Ohio.

Website: https://cfa.org

FB

Felis Britannica

The English branch of the Luxembourg based association was founded in 1949 in Paris, France.

Website: https://felis-britannica.org.uk

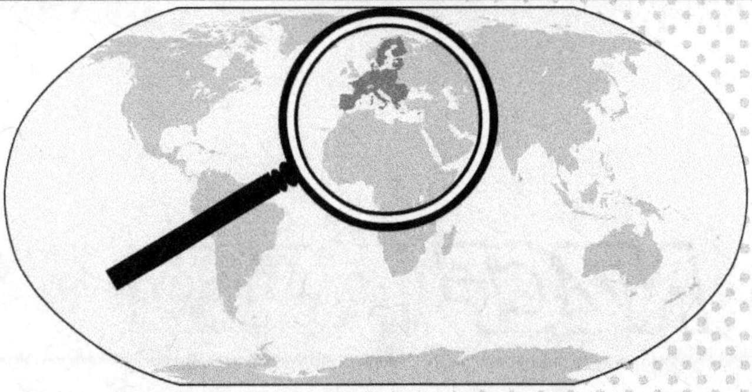

FIFe

Fédération Internationale

The association was founded in 1949 in

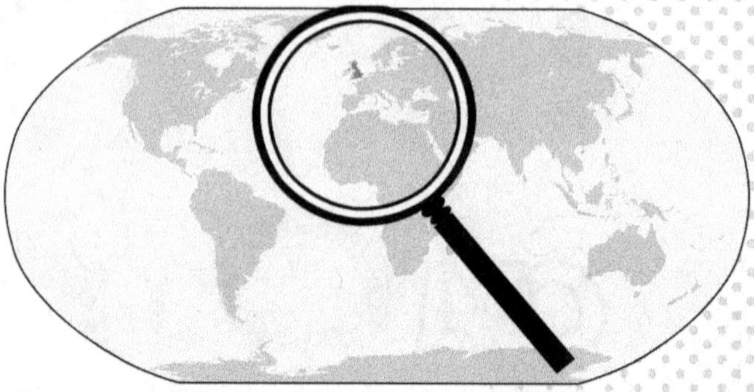

GCCF

Governing Council of the Cat Fancy

The association was founded in England in 1887. It is based in Bridgwater, Somerset.

Website: https://gccfcats.org

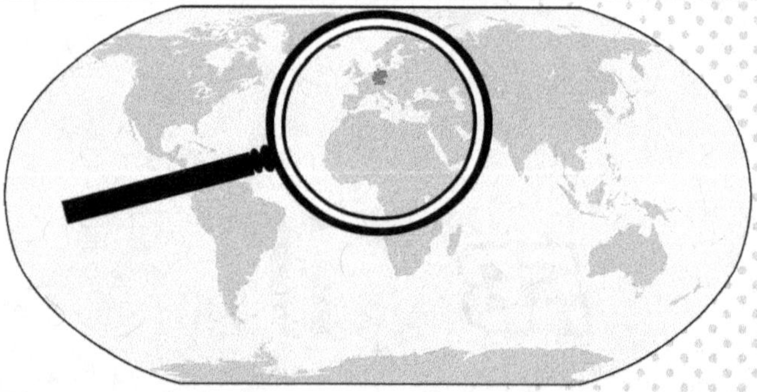

LOOF

Fédération pour la gestion du Livre Officiel des Origines

The association was founded in France in 1996 and is based in Paris.

Website: https://loof.asso.fr

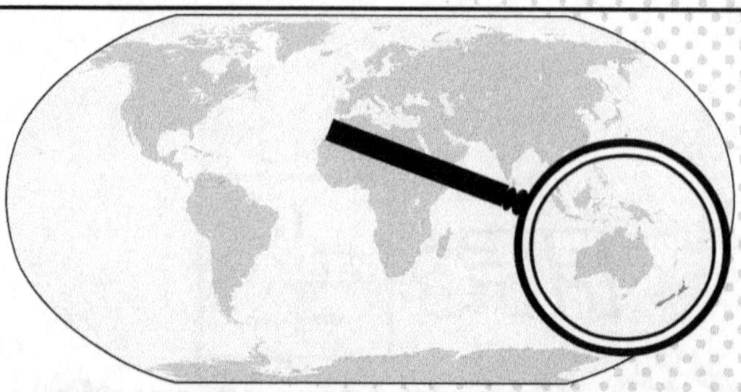

NZCF

New Zealand cat Fancy

The New Zealand branch of the English GCCF was established in the 1930s and is based in Te Puke, Auckland.

Website: https://nzcf.com

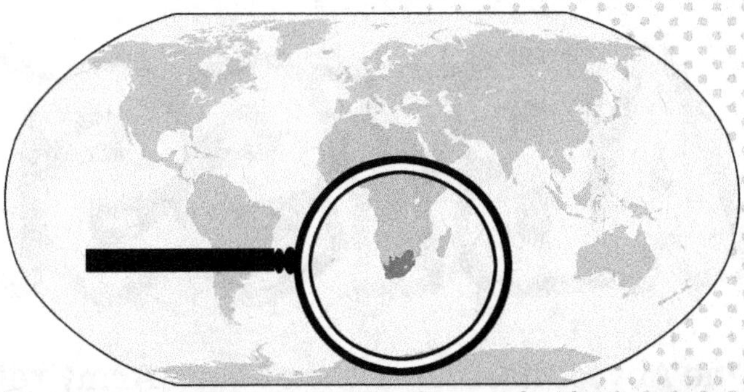

SACC

Southern Africa cat Council

The association was founded in 1945 and is based in Kensington, Gauteng.

Website: https://tsacc.org.za

TICA

The International cat Association

The world's largest pedigree cat registration was founded in 1979 in Branson, Missouri.

Website: https://tica.org

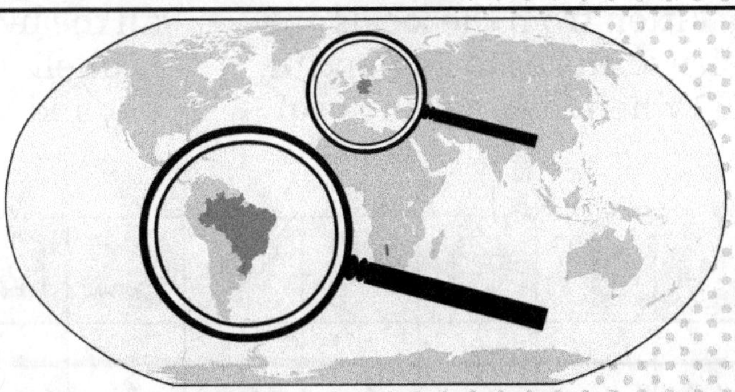

WCF

The association was founded in 1988

Breeding start 1868 United Kingdom

TOP 10

ABYSSINIAN

The Abyssinian cat is one of the oldest cats in the world. It was trendy in ancient Egypt and was deified. For a long time, it has been assumed that this elegant, muscular cat, whose appearance is strongly reminiscent of a miniature puma, comes from the East African Empire of Abyssinia itself - today's Ethiopia. But genetic research has shown that the pretty cat comes from the South Asian jungle and must have been brought to the Kingdom of Abyssinia via trade routes of the Vikings, Arabs, Turks, and British. The first documented Abyssinian cat was called Zula. She lived with the wife of a British army captain who bought the kitten from the locals and took it with her to England in 1868 when the English Royal Army returned. In England, the cat was received with enthusiasm because of its beautiful fur. The coat consisted of an embroidered tabby pattern, which essentially only appeared in wild animals, such as field hares. The tabby pattern seems to be monochrome but has a remarkable play of colors, starting from the M-shaped forehead drawing. The so-called agouti gene causes each hair of the Abyssinian cat to grow with horizontal stripes. The ticked hairs are light and dark banded and end at a dark tip. The quiet and, therefore, all the livelier Abyssinian cat is intelligent, curious, sociable, and human-oriented.

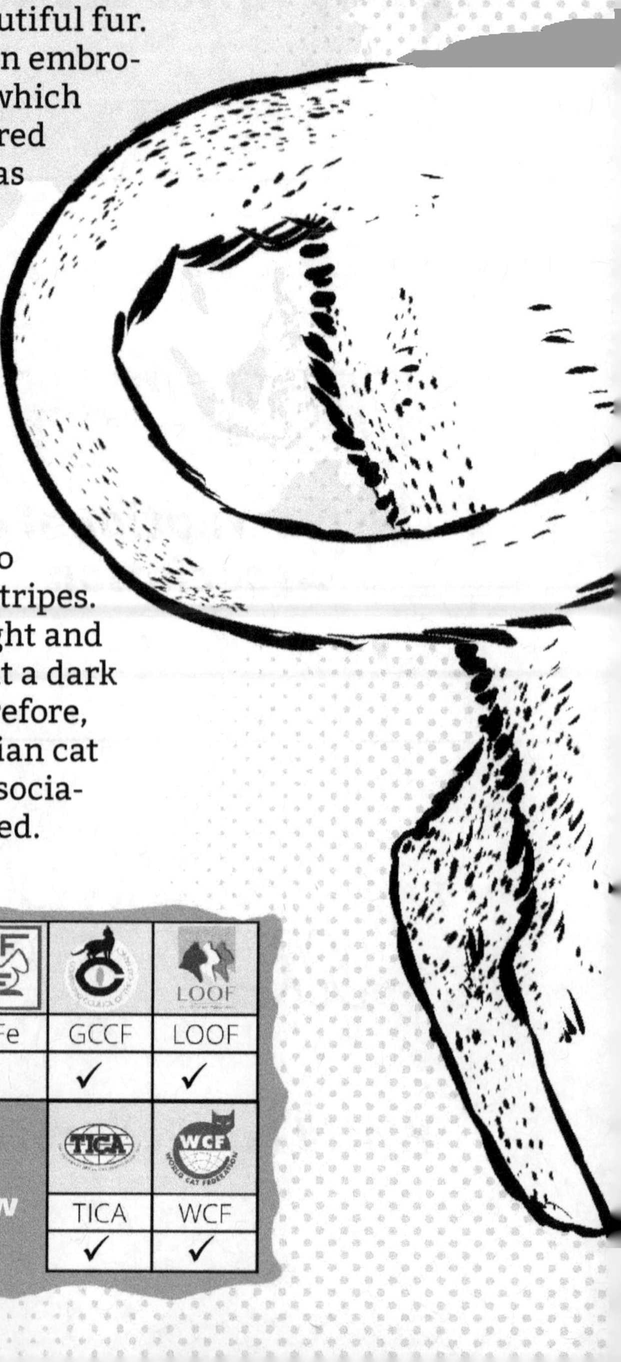

ACF	ACFA	CAA	CCA-AFC	CCC	CCCA	CFA	FB	FIFe	GCCF	LOOF
✓	✓	✓	✓	✓	✓	✓	✓	✓	✓	✓

NZCF	SACC	TICA	WCF
✓	✓	✓	✓

Selection of the most important cat associations in the world: Australia, Canada, China, France, Germany, New Zealand, South Africa, UK and USA

Recognized since 1882

LIFE EXPECTATION
9-15 YEARS

The Cat Fanciers Association (CFA) and The International Cat Association (TICA) rank the Abyssinian cat among the top 5 most popular cat breeds.

Market price:
200.00 USD–
4,000.00 USD
(2025)

AMERICAN BURMESE

The Burma cat or Burmese is a very old cat breed and comes from temples in Burma, today's Myanmar. It is said to have been kept in the 15th century by monks as one of 16 cat breeds. She was very revered and called Maeo Thong Daeng (copper cat) or Suphalak (auspicious character). Her younger breeding history took place as follows: In 1933, US Navy Doctor Joseph C. Thompson flew back from Rangoon to San Francisco. He was accompanied by the Burmese female Wong Mau - to date the first and then only Burmese cat to enter US soil. Dr. Thompson was a psychiatrist and often allowed her to participate in his therapy sessions. One may puzzle over whether Wong Mau may have interfered too often in the sessions or talked too much in principle - at any rate, it was obviously Thompson too much at some point, and he turned away from the cat lady. He started the first comprehensive breeding program in the history of a cat breeding with a group of geneticists and

ACF	ACFA	CAA	CCA-AFC	CCC	CCCA	CFA	FB	FIFe	GCCF	LOOF
✓	✓	✓	✓	✓	✓	✓	✓	✓	✓	

NZCF	SACC	TICA	WCF
✓	✓	✓	✓

Selection of the most important cat associations in the world: Australia, Canada, China, France, Germany, New Zealand, South Africa, UK and USA

ORIGIN: SOUTHEAST ASIA, UNITED STATES

Recognized since 2000

breeders. Wong Mau was paired with the Seal point Siam male Tai Mau and then with her son when he reached sexual maturity. The revenge act of Dr. Thompson formed the Foundation of the modern Burma cat. In 1936 the Burmese cat was officially recognized as a cat breed. But the British weren't idle either. In the course of their Victorian colonial rule, they had already brought Burmese cats into the country much earlier: in 1871, a cat exhibition in the Chrystal Palace showed a cat that - in retrospect - resembled Dr. Thompson's Burmese cat in every way. Within American Burmese cats, there are two variations: the traditional and the contemporary. The latter has larger eyes and a pronounced nose - which is a doll-like facial expression. American Burma cats are only accepted in the colors sable (brown), champagne, blue, and platinum.

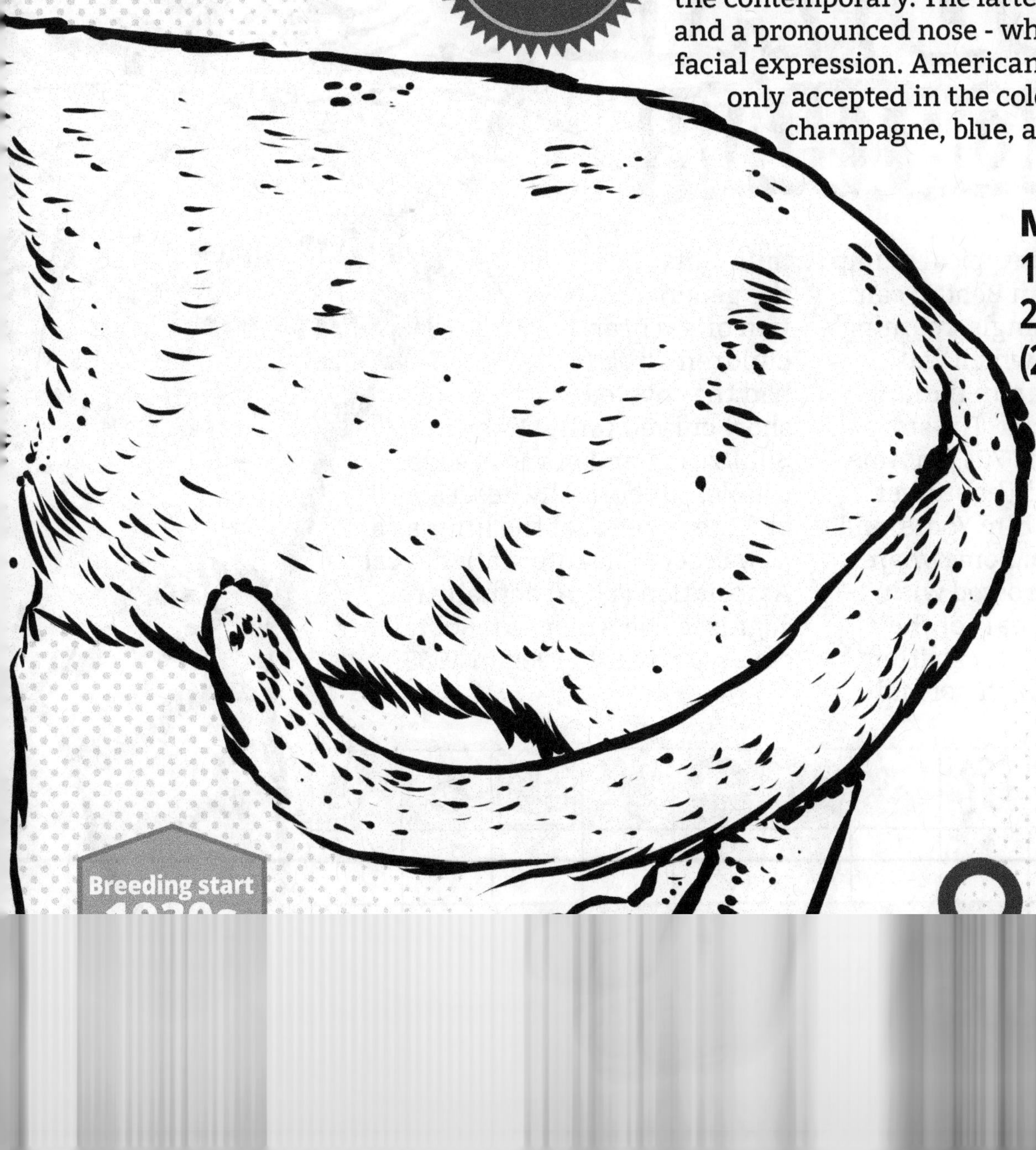

Market price:
150.00 USD–
2,800.00 USD
(2025)

Burma head defect (BHD), diabetes mellitus, hypokalemia, pica, gangliosidosis, flat-chested kitten syndrome (FCKS), health problems with eyes, kidneys (stones) and sensory organs

Breeding start

AMERICAN BOBTAIL
SHORTHAIR

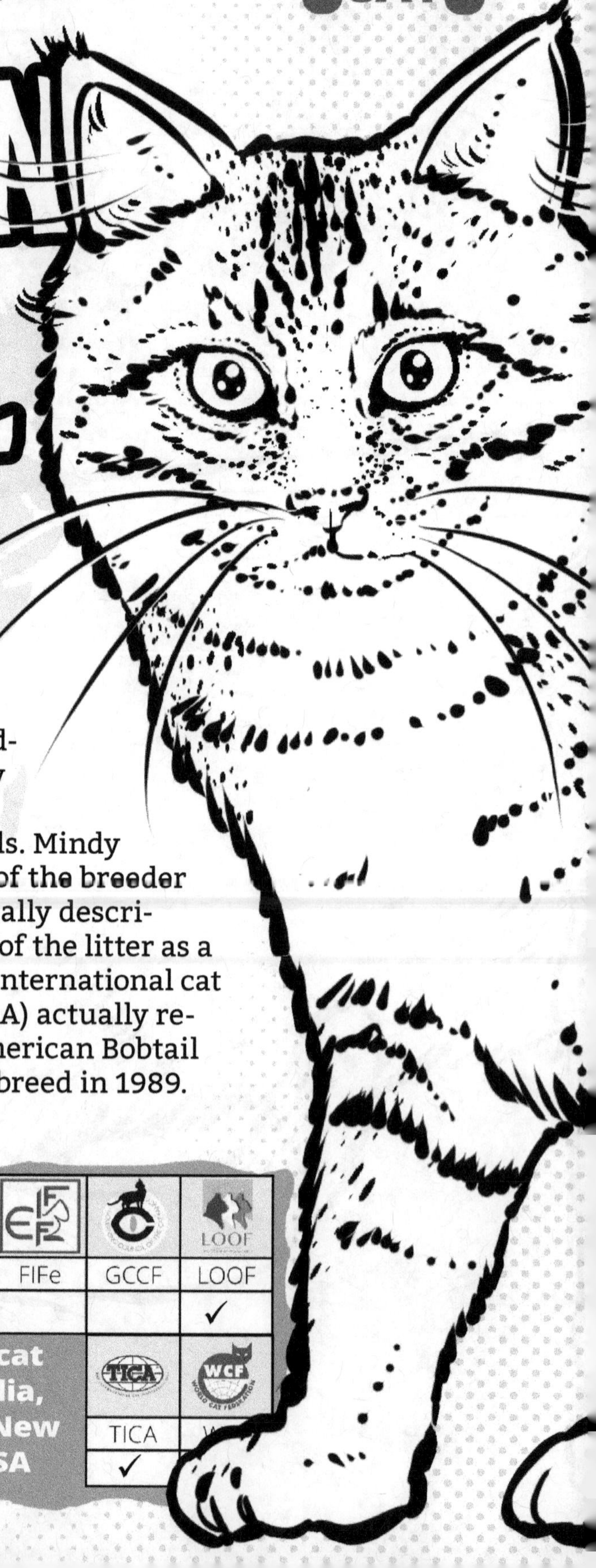

Recognizable by its typical stump tail, the American Bobtail can be traced back to a single, roaming male with tiger fur, which was found in the 60s near an Indian reservation in Arizona. The story goes so far that in the 70s, an American couple adopted the strayer on vacation, named him Yodie, and brought him to Iowa. Somewhere in Iowa, Yodie was crossed with a Siamese female as part of the breeding program. In their litter, there were puppies with normal tails. But the generation of the grandchildren already had the coveted, short curved tails. Mindy Shultz, a friend of the breeder couple, euphorically described the puppies of the litter as a new breed. The International cat Association (TICA) actually recognized the American Bobtail cat as a new cat breed in 1989.

ACF	ACFA	CAA	CCA-AFC	CCC	CCCA	CFA	FB	FIFe	GCCF	LOOF
	✓	✓				✓				✓

NZCF	SACC	TICA	W
		✓	

Selection of the most important cat associations in the world: Australia, Canada, China, France, Germany, New Zealand, South Africa, UK and USA

ATELIER KAYMAK®

Science PAPERS

An American Bobtail cat needs between 2 and 3 years to develop fully.

Recognized since 1882

Market price:
150.00 USD–1,500.00 USD (2025)

Breeding start 1960s United States

AMERICAN BOBTAIL LONGHAIR

The Longhair variant of the American Stumpy tail cat was created by mixing a short-haired American Bobtail cat with a Himalayan cat. Its coat is shaggy rather than fluffy and gives the cat a wild look reminiscent of its truly wild cousin, the North American lynx. Although some origin stories point to a cross between a wild red lynx and a domestic cat in the 1970s, this seems very unlikely. The famous short tails of Bobtail cats are bushy, can be straight or wavy, and vary from non-existent (Rumpy), too small (Stumpy), and long. American stumpy-tailed cats may not be as agile as cats with standard length, curved tails - but they are calmer, friendlier, and more sociable.

ACF	ACFA	CAA	CCA-AFC	CCC	CCCA	CFA	FB	FIFe	GCCF	LOOF
	✓	✓				✓				✓

NZCF	SACC	TICA	WCF
		✓	

Selection of the most important cat associations in the world: Australia, Canada, China, France, Germany, New Zealand, South Africa, UK and USA

ATELIER KAYMAK
Science PAPERS

Breeding start 1960s United States

BOBTAIL CAT

Recognized since 1989

BIG CAT

LIFE EXPECTATION 11-15 YEARS

Market price:
150.00 USD–
1,500.00 USD
(2025)

AMERICAN CURL SHORTHAIR

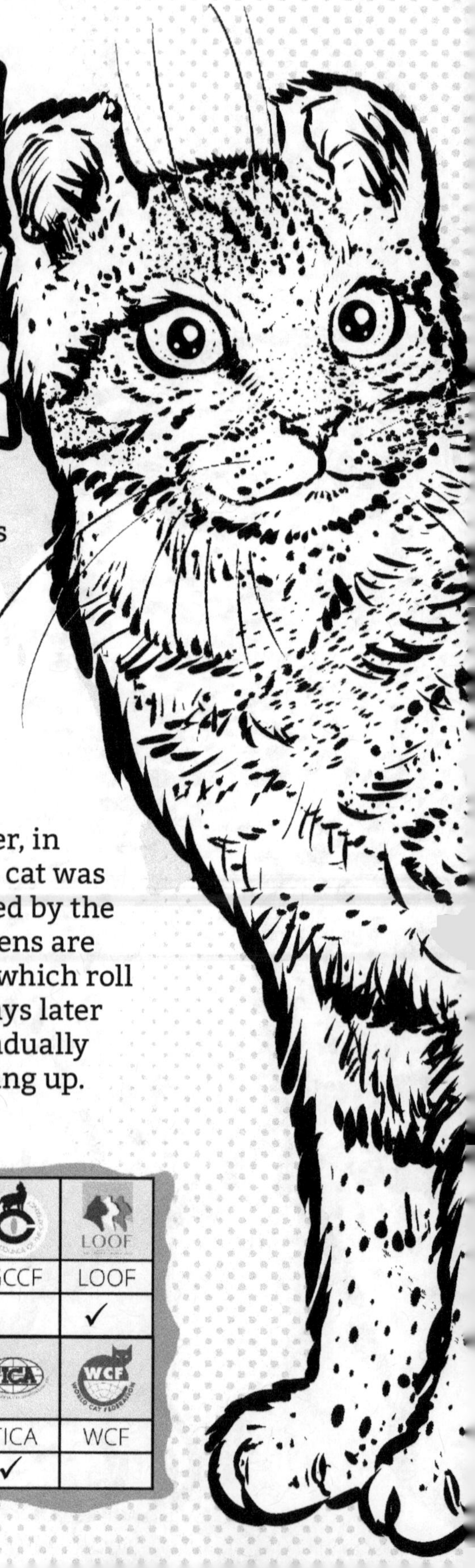

The American Curl cat is a relatively new breed that was discovered by chance. In 1981, the couple Grace and John Ruga found two stray kittens on their doorstep in Lakewood, California. A street tomcat and a pretty pregnant female cat with strangely shaped ears. Grace and John immediately took care of the two cats and gave them a new home. Swaying into safety, the female cat, now called Shulamith, gave birth to healthy offspring.
In this litter, there were beautiful looking little kittens with the same distinctive, short, and curved ears as the pretty mother. The couple Ruga immediately agreed to breed these strange, lovely cats and make them accessible to other people.
In 1983 they presented two of their cats for the first time at a professional cat show in Palm Springs, where they caused a sensation. Two years later, in 1985, the American Curl cat was recognized as a new breed by the American TICA. The kittens are born with straight ears, which roll up closely again a few days later - only to unroll again gradually during the time of growing up.

ACF	ACFA	CAA	CCA-AFC	CCC	CCCA	CFA	FB	FIFe	GCCF	LOOF
	✓	✓				✓				✓

NZCF	SACC	TICA	WCF
		✓	

Selection of the most important cat associations in the world: Australia, Canada, China, France, Germany, New Zealand, South Africa, UK and USA

ATELIER KAYMAK

Science PAPERS

The American Curl cat has large circular eyes and a great tail, which reminds us of a feather duster not only because of its mobility.

Market price:
$100 to $3,000 (2023)

LUXUS

Recognized since 1986

Breeding start
1981
United States

LIFE EXPECTATION
12-16 YEARS

AMERICAN CURL LONGHAIR

All American Curl cats living today can trace their pedigree back to Grace and John Rugas Shulamith. The Curl gene is inherited autosomal dominant, which means that each kitten whose both parents are American Curl Cats - has the gene mutation. The American Curl cat is medium-sized, has a roundish head with a pronounced muzzle. It has large, friendly almond eyes and the said uniquely curled hind ears with horizontally protruding tufted hair. Her mischievous to cheeky-looking facial expression immediately wins the hearts of all cat lovers. The sociable American Curl cat remains youthful well into adulthood: it is active, curious, and playful. She gets along particularly well with children and is therefore suitable without restrictions as a family animal.

ACF	ACFA	CAA	CCA-AFC	CCC	CCCA	CFA	FB	FIFe	GCCF	LOOF
✓	✓	✓		✓	✓	✓	✓	✓		✓

NZCF	SACC	TICA	WCF
		✓	✓

Selection of the most important cat associations in the world: Australia, Canada, China, France, Germany, New Zealand, South Africa, UK and USA

ORIGIN:

UNITED STATES

Breeding start
1981
United
States

LIFE
EXPECTATION
12-16
YEARS

Recognized
since
1992

LUXUS

Market price:
200.00 USD–
3,500.00 USD
(2025)

Progressive Retinal atrophy (PTR)

AMERICAN SHORTHAIR

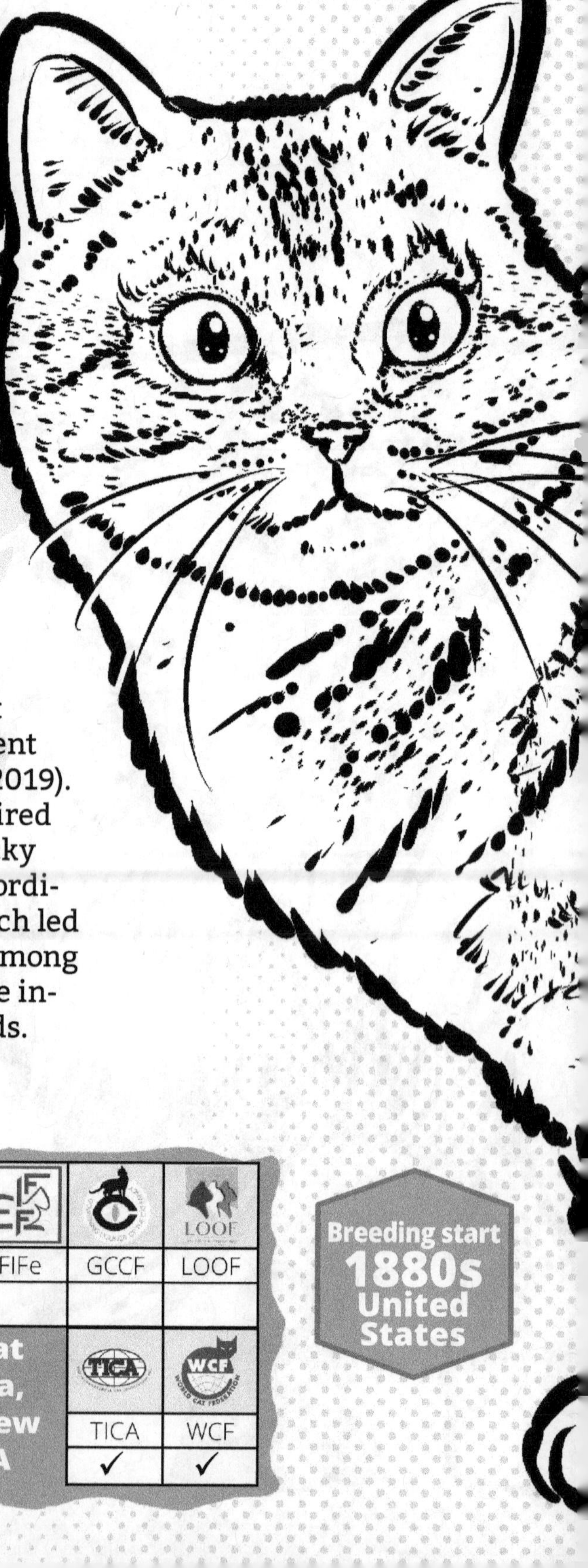

The history of this race traces the birth of a nation. Legend says that in the 17th century, the ancestors of these cats accompanied the Pilgrim Fathers aboard their ships - including the Mayflower - to the New World. The cats were there when the Separatists, Congregationalists, and Anglican emigrants who had fled exiled from England founded the Plymouth Colony in New England (now Massachusetts). The cats helped the settlers protect the food provided by the surrounding Indian tribes from vermin. In the 19th century, their services as pied pipers were in great demand in the California Gold Rush mines. A cat cost up to 50 dollars each at that time - the equivalent of 1850 dollars today (2019). The American shorthaired cat is compact and stocky and thus more like an ordinary domestic cat, which led to a certain snobbery among cat lovers who are more interested in exotic breeds.

ACF	ACFA	CAA	CCA-AFC	CCC	CCCA	CFA	FB	FIFe	GCCF	LOOF
✓	✓	✓	✓	✓	✓	✓				

NZCF	SACC	TICA	WCF
	✓	✓	✓

Selection of the most important cat associations in the world: Australia, Canada, China, France, Germany, New Zealand, South Africa, UK and USA

Breeding start
1880s
United States

ATELIER KAYMAK

Science PAPERS

ORIGIN:

EUROPE, UNITED STATES

Thanks to patient breeding, the former Domestic Shorthair cat made a name for itself on the cat breeders' Catwalk in the USA and was ennobled in 1966 with the new name American Shorthair cat.

Recognized since 1966

Market price:
100.00 USD–2,000.00 USD (2025)

LIFE EXPECTATION 15-17 YEARS

(with pure white coat): congenital hearing

AMERICAN TIFFANY

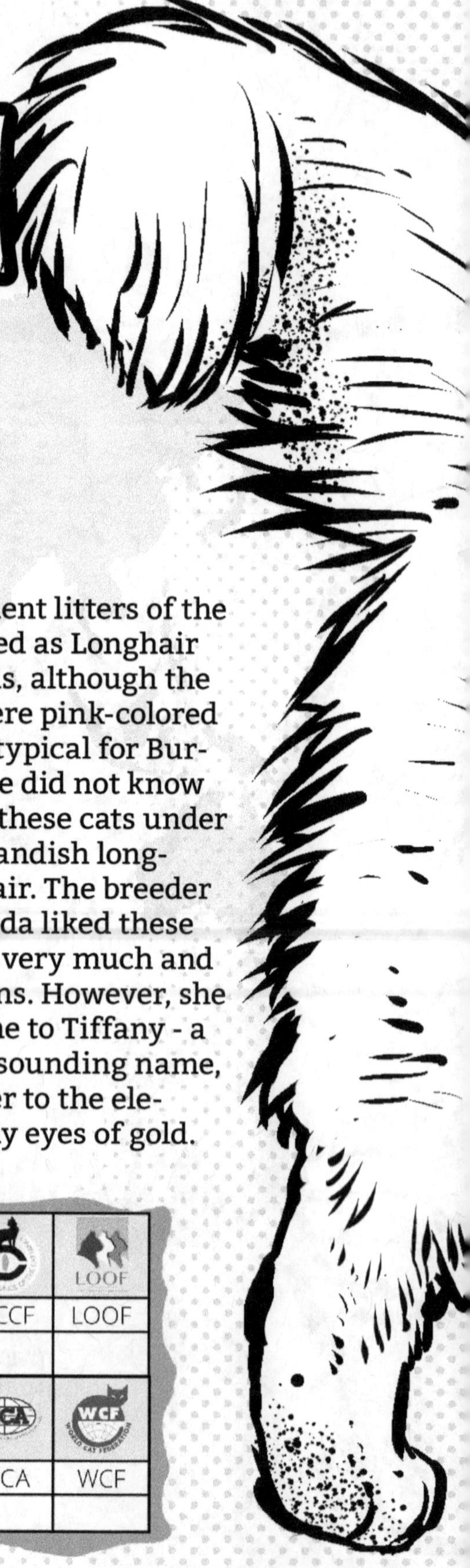

American Tiffany cats or Chantilly-Tiffany cats, as they are also called, were initially only accepted in chocolate colors and with lush fur. Today their color varies from blue to lilac to tabby. Their eyes are often yellow. With increasing age, they become golden. Their owners often affectionately refer to them as honey eyes. The origin of the cat breed is somewhat mysterious. In 1967 a stranger allegedly bought two cats from the New Yorker Jennie Robinson and started breeding. Both cats had a fluffy chocolate coat - but there was no other exact evidence of the breed, and the origin around these cats remained in the dark. The kittens in the subsequent litters of the two were first classified as Longhair Burmese cats. And this, although the paws of the kittens were pink-colored and not in the brown typical for Burmese cats. Because one did not know the origin, one seized these cats under the simple name Outlandish longhair - ...Foreign longhair. The breeder Sigyn Lund from Florida liked these strange Longhair cats very much and bought some specimens. However, she changed her ugly name to Tiffany - a much more beautiful sounding name, which also fitted better to the elegant coat and the shiny eyes of gold.

ACF	ACFA	CAA	CCA-AFC	CCC	CCCA	CFA	FB	FIFe	GCCF	LOOF
			✓							

NZCF	SACC	TICA	WCF

Selection of the most important cat associations in the world: Australia, Canada, China, France, Germany, New Zealand, South Africa, UK and USA

The name caused confusion because of the British Tiffanie cat of the same name. Therefore some cat federations saw themselves forced to a further name change and called the Tiffany from then on Chantilly. To remember this, the double name Chantilly-Tiffany is often used for this reason.

Recognized since 1970

Market price:
400.00 USD–1,200.00 USD (2023)

LIFE EXPECTATION
14-16 YEARS

Breeding start
1967
United States

Breeding start
1966
United States

AMERICAN WIREHAIR

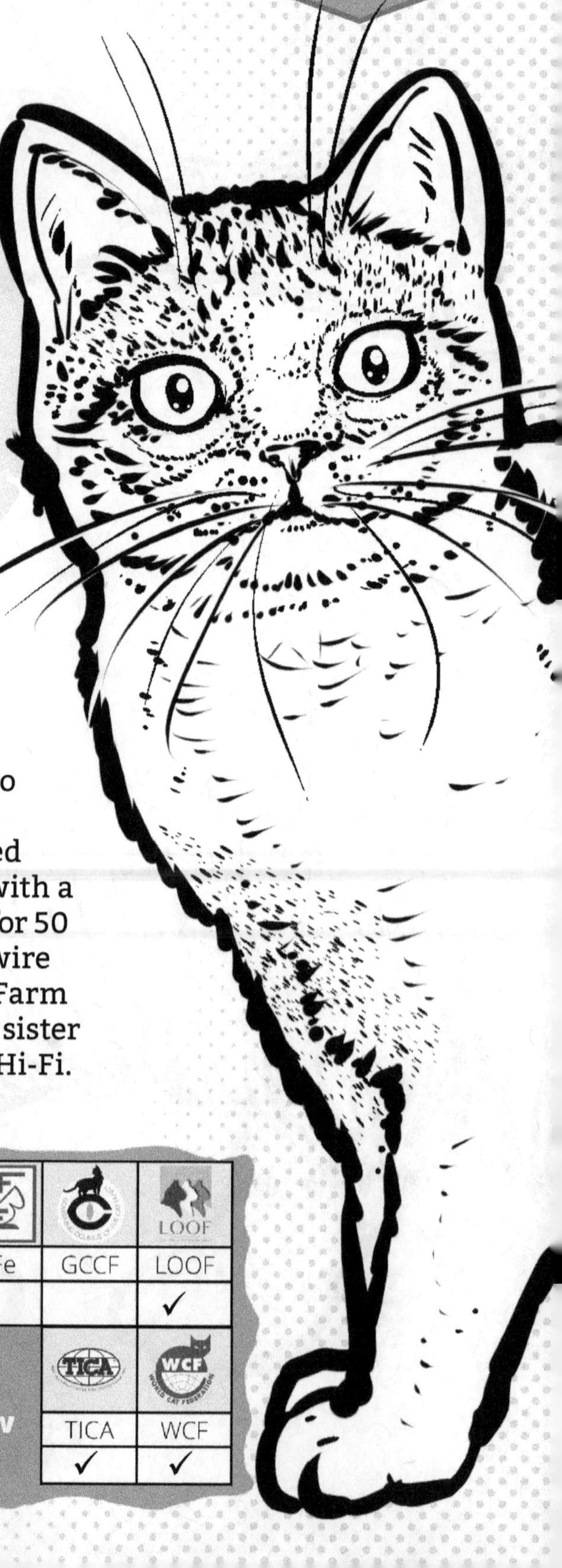

In 1966, farmers in the hinterland of New York discovered something strange in their cat's litter: among the six newborn puppies were two rather strange-looking kittens with disheveled fur. The farmers had never seen anything like it on cats before. The animals looked as if they had received an electric shock. In fact, each hair of the fur was undulated and curled at the end. Even the beard hairs of the poor cats were bent. The comparison to living wire hairbrushes was evident, and so the American Wirehair cat got its name. The wire hair effect is a rare mutation in cats caused by the gene of the same name. The cat breeder Joan O'Shea, who specializes in Rex cats, acquired the two mutated kittens in combination with a normal-looking sibling for 50 dollars. She named the wire hair male Council Rock Farm Adam of Hi-Fi. The little sister got the name Tip-Toe of Hi-Fi.

ACF	ACFA	CAA	CCA-AFC	CCC	CCCA	CFA	FB	FIFe	GCCF	LOOF
	✓	✓	✓	✓		✓				✓

NZCF	SACC	TICA	WCF
		✓	✓

Selection of the most important cat associations in the world: Australia, Canada, China, France, Germany, New Zealand, South Africa, UK and USA

LIFE EXPECTATION
14-18 YEARS

Probably she thought of the Bible, Frankenstein, and the effects of loud heavy metal music from very, very large loudspeakers on the psyche and fur of the cats. Due to its rarity, the American Wire Hair cat was quickly sought after by breeders all over the world. Although the American Wire Hair cat was registered in 1967, it was not officially recognized by the CFA until ten years later.

Recognized since 1977

Market price:
150.00 USD–1,200.00 USD (2023)

(with pure white coat): congenital hearing loss, deafness, disturbances of the sense of balance, Eye tremor, Progressive Retinal atrophy (PTR)

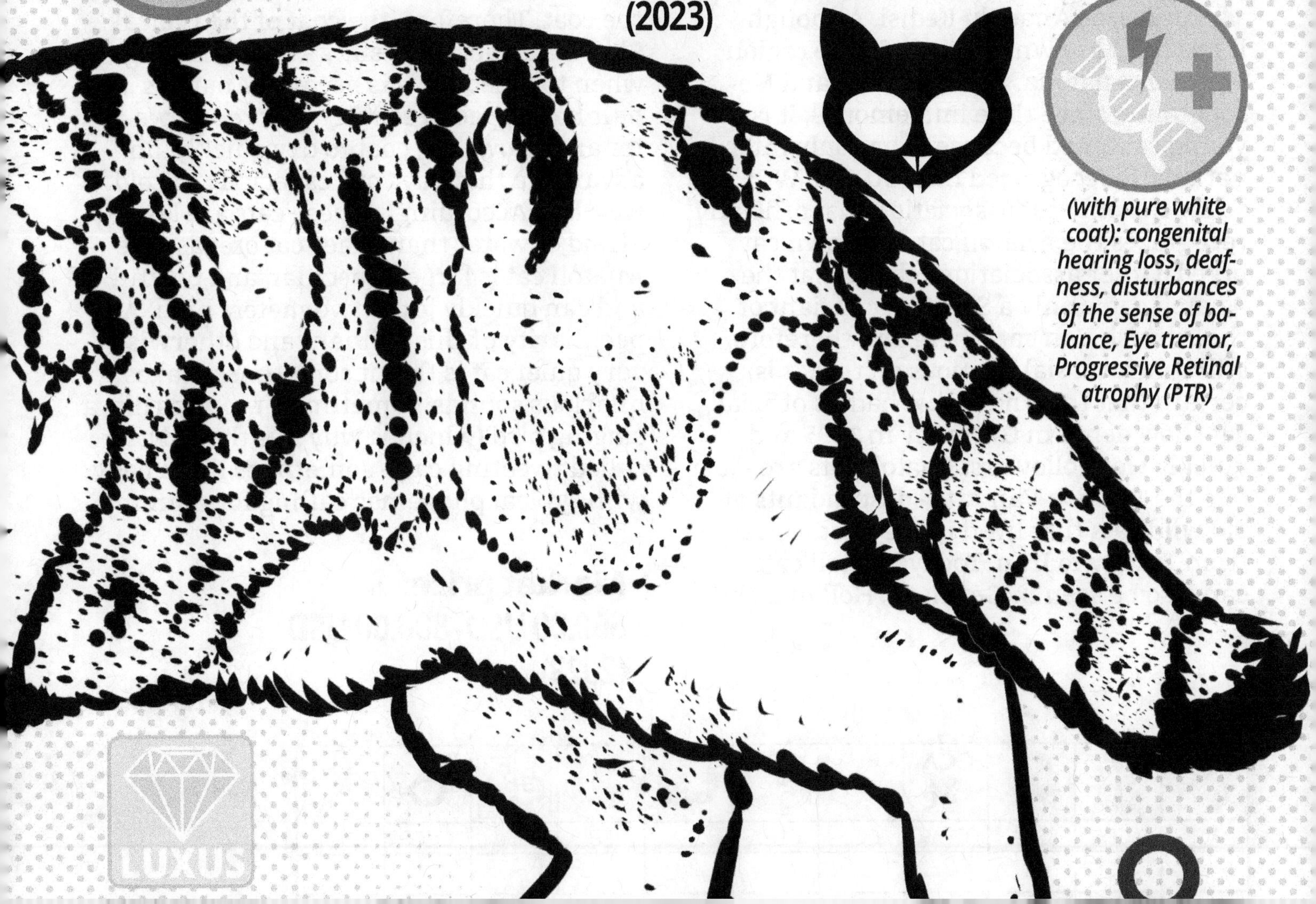

LUXUS

SHORT HAIR

Breeding start
1995
Germany

ANATOLI

The Anatoli cat is known in its homeland, Turkey, as Anadolu Kedisi. Although it has been known to people in the region between Anatolia, the Caucasus, and Mesopotamia since time immemorial, it is a young cat breed because it was only internationally recognized in 2000. The WCF is so far the only cat association that officially recognizes the Anatoli cat as its own cat breed. Other associations claim that the Anatoli cat is only a Shorthair variant of the Turkish Van cat. There are, therefore, still controversial opinions on this. It is certain that the Anatoli cat had its official cat show debut in Germany in 1995 and quickly won followers. Anatoli cats are - just like the Turkish Van cats - descendants of old wild cats that lived in the Middle East from Turkey to Iran. Today Anatoli cats can also be seen in Germany, Holland, and Russia. The Anatoli has no undercoat in the coat. Therefore, the coat of the typical Shorthair cats looks silky - but feels rough when touched. This is because Anatoli's hair has a special coating that repels water and allows the coat to dry quickly, an advantage that pays off, especially in cold weather. Accordingly, these cats are less afraid of water than other cat breeds. The Anatoli cat is large, muscular, and playful and can quickly jump two meters high. A peculiarity of this brilliant and otherwise very quiet cat is that it sometimes makes hilarious sounds: sometimes it sounds like a singing bird (which is why it is the only cat called tweeting cat) - sometimes like a baby with typical phonemes calling for mummy.

Market price:
500.00 USD–800.00 USD (2023)

LIFE EXPECTATION
15-20
YEARS

ACF	ACFA	CAA	CCA-AFC	CCC	CCCA	CFA	FB	FIFe	GCCF	LOOF

NZCF	SACC	TICA	WCF
			✓

Selection of the most important cat associations in the world: Australia, Canada, China, France, Germany, New Zealand, South Africa, UK and USA

ORIGIN:
TURKEY, GERMANY
Recognized since 2000

SHORT HAIR

BIG CAT

APHRODITE SHORTHAIR

The cat breed originated on the island of Cyprus as a result of natural reproduction and not breeding. The Aphrodite is a genuine survivor: it defies the harsh winters of the highlands and survives any hot, dry summer. This is probably due to natural selection, which only allowed the strongest cats to develop the necessary resistance in their struggle for survival against nature. The free-living Aphrodite lives, like many of its cousins, the wild cats, from prey such as rats, snakes, and other reptiles. The Anglo-Saxon name of the Aphrodite is translated as Cypriot Mountain Giant cat or Aphrodite's Giant - and probably expresses better what kind of cat it is. The Aphrodite is, as the name suggests, a very big cat. In fact, the remarkable Aphrodite is one of the biggest cats in the world! She towers over domestic cats by a lot and makes them look like little toy kittens compared to her. The Aphrodite cat has a broad, muscular physique, which is already noticeable as a kitten and attracts amazed glances. In 2012, the World Cat Federation (WCF) became the first federation to take notice of this remarkable cat. The following year, 2017, the International Cat Association (TICA) recognized her as a temporary new breed. TICA claims that the Aphrodite shows downright dog-like behavior when affectionate to her owners.

Breeding start
2006
Cyprus

Market price:
100.00 USD–
800.00 USD
(2023)

LIFE EXPECTATION
12-15
YEARS

ACF	ACFA	CAA	CCA-AFC	CCC	CCCA	CFA	FB	FIFe	GCCF	LOOF
✓										

NZCF	SACC	TICA	WCF
			✓

Selection of the most important cat associations in the world: Australia, Canada, China, France, Germany, New Zealand, South Africa, UK and USA

Weight
10-18
lbs

ORIGIN:
CYPRUS, EGYPT, MIDDLE EAST
Recognized since 2012

APHRODITE LONGHAIR

Surrounded by the three ancient cradles of civilization - Greece, Mesopotamia, and Egypt - Cyprus has a long and mythical past. The same is true for Cypriot cats! Archaeological funds from a Neolithic tomb around 9500 B. C. provide clear evidence of human-cat coexistence. In which way the cats reach the island during the Stone Age can only be guessed. With high probability, however, both the Aphrodite cat and the modern Cyprus cat or St Helena cat - do not descend directly from these Stone Age cats - but were introduced to the island much later. The official breeding program with Aphrodite cats started in 2006 and was developed by Walter and Theresa Litherland. The Aphrodite cat is, as already mentioned, very big and muscular - but not full or even fat. She has a friendly nature and likes the company of people.

Breeding start 2006 Cyprus

BIG CAT

Market price: 100.00 USD–800.00 USD (2023)

LIFE EXPECTATION 12-15 YEARS

Recognized since 2012

ACF	ACFA	CAA	CCA-AFC	CCC	CCCA	CFA	FB	FIFe	GCCF	LOOF
✓										

NZCF	SACC	TICA	WCF

Selection of the most important cat associations in the world: Australia, Canada, China, France, Germany, New Zealand, South Africa, UK and USA

Weight 10-18 lbs

ORIGIN:
CYPRUS, EGYPT, MIDDLE EAST

ARABIAN MAU

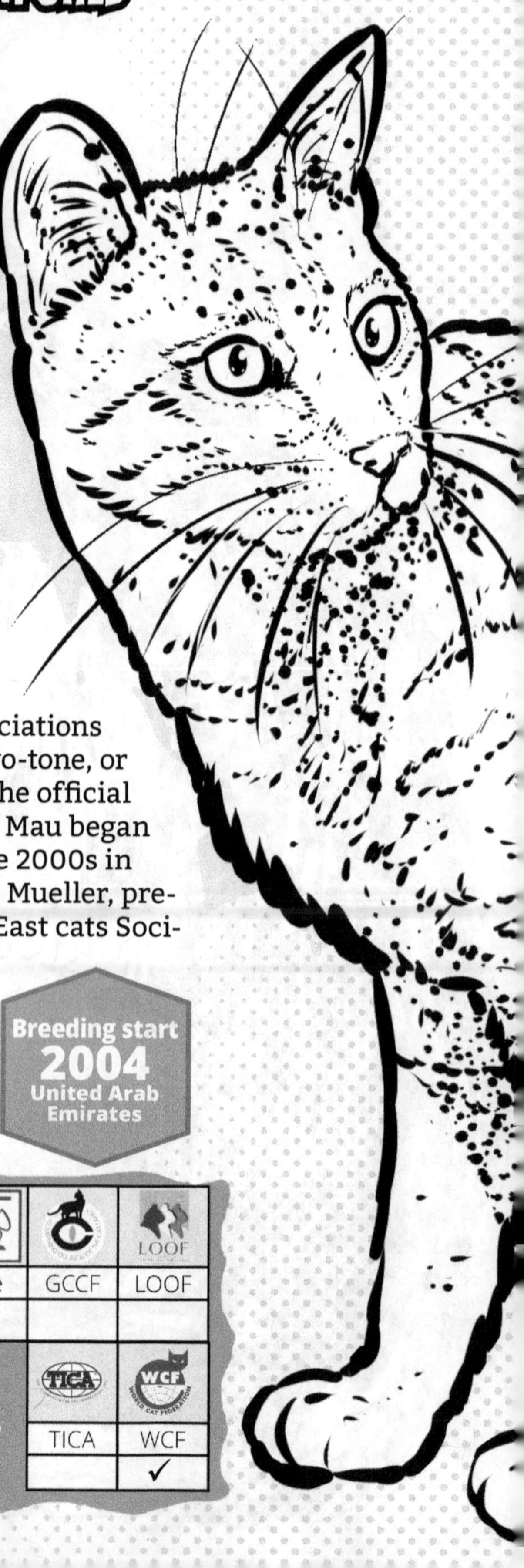

The Arabian Mau cat is descended from desert cats that arrived on the Arabian Peninsula more than a millennium ago. It is, therefore, now a native, naturally occurring cat that lives together with the locals in the dusty, dry cities of Arabia. The robust Arabic Mau-Cat is slim and has long, muscular legs and a pointed tail. The male is much more muscular and fleshier than the female and therefore appears much larger. The Arabic Mau cats are very reserved, shy, and distant. She is a balanced, talented hunter whose genes carry the survival mechanisms for living in extreme desert conditions. She is recognized by the cat associations as having a tabby, two-tone, or white coat pattern. The official breeding of the Arab Mau began in the first half of the 2000s in Dubai. Petra-Khadija Mueller, president of the Middle East cats Society, is said to have kept over 50 Arab Mau cats in her home at the time.

Breeding start
2004
United Arab Emirates

ACF	ACFA	CAA	CCA-AFC	CCC	CCCA	CFA	FB	FIFe	GCCF	LOOF

NZCF	SACC	TICA	WCF
			✓

Selection of the most important cat associations in the world: Australia, Canada, China, France, Germany, New Zealand, South Africa, UK and USA

ATELIER KAYMAK
Science PAPERS

The Arab Mau cat was recognized as a new cat breed by the WCF in 2008.

Market price:
500.00 USD–800.00 USD (2023)

LIFE EXPECTATION
12-14 YEARS

Recognized since 2008

ANCIENT ORIGIN

ASIAN ALSO MALAYAN

All Asian cats and Malayan cats are traced back to one litter of cats by Miranda Bickford- Smith. The four kittens, which were born in 1981 due to an unintentional crossing of two breeding cats, are the progenitors of all Asian cats living today. The elegant Asian cat has the same temperament as the active Burmese cat. The small, dainty legs do not inhibit her play instinct in any way, so you can see her like a fur ball all the time running through the area. The Asian cat has a wedge-shaped face with pronounced cheekbones, which gives it an attractive appearance. Her eye color varies from yellow to amber. Although the Asian cat is somewhat less playful than her sister, the Burma cat, she is just as loving and chatty.

ACF	ACFA	CAA	CCA-AFC	CCC	CCCA	CFA	FB	FIFe	GCCF	LOOF
									✓	

NZCF	SACC	TICA	WCF
			✓

Selection of the most important cat associations in the world: Australia, Canada, China, France, Germany, New Zealand, South Africa, UK and USA

Her short-haired coat is shiny and can show all kinds of colors. The Asian cat is one of the first cat breeds in Great Britain to have its friendly nature assigned as an integral part of the scoring standard. It is through her that the saying If it is not beautiful, it is not Malayan came into being.

Market price:
200.00 USD–
4,500.00 USD
(2025)

Breeding start
1981
United Kingdom

Recognized since
2000

LIFE EXPECTATION
12-15
YEARS

Weight
9-16
lbs

AUSTRALIAN MIST

The Australian Mist is the breeding of Dr. Truda Straede in Australia, which was developed in 1976. The medium-sized and temperamental cat with the distinctive coat pattern is a cross between an Abyssinian cat (50%), a Burmese cat (25%), and a domestic cat (25%). The coat of the beautiful cat is monochrome greyish to cloudy white on the back and runs as a tabby pattern over the belly side, which led to the earlier naming - Spotted Mist. Although the marbled pattern was also accepted in 1997, this led to the name being changed to Australian Mist the following year. Although one could assume that this rather curious and playful cat is an unconditional free goer - it is very eager for the proximity of humans. She is even said to prefer the company of humans more than nature. She is, therefore, an excellent house-cat.

ACF	ACFA	CAA	CCA-AFC	CCC	CCCA	CFA	FB	FIFe	GCCF	LOOF
✓	✓	✓			✓				✓	

NZCF	SACC	TICA	WCF
✓		✓	✓

Selection of the most important cat associations in the world: Australia, Canada, China, France, Germany, New Zealand, South Africa, UK and USA

ATELIER KAYMAK
Science
PAPERS

Australian Mist cats are known to be very calm and patient and only irritable in rare exceptional cases. Thus, they are suitable outstanding for the attitude in child-rich families.

Recognized since 1882

Market price:
100.00 USD–1,200.00 USD (2023)

Breeding start 1880s United States

Hypokalaemia (potassium deficiency)

LIFE EXPECTATION 14-16 YEARS

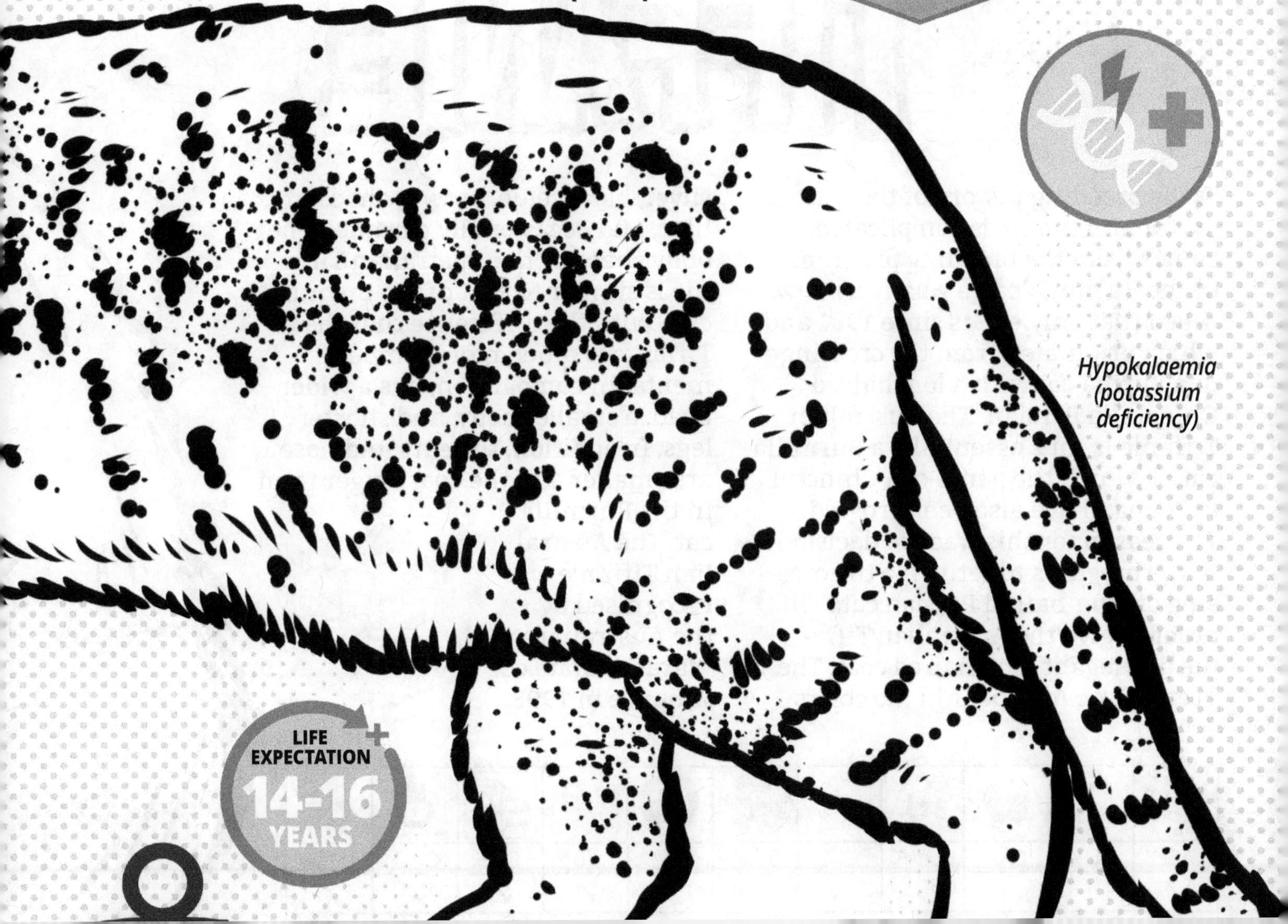

AUSTRALIAN TIFFANIE

The breeding history of the Australian Tiffanie is complicated. It comes from the breeding program of the Burmilla cat, a relatively new breed that only exists since 1981 and which originated from the crossing of a Burma cat with a longhaired Chinchilla Persian. The Australian Tiffanie is thus essentially a Burmilla cat, to which the genes of a Chinchilla Persian have also been crossed. The reason for this was the decision of the breeders to return to the crossing of long-haired Persian cats - in order to give the Australian Tiffanie an elegant long-haired coat. The pretty fur can appear in the colors silver, black, blue, brown, chocolate, lilac, and occasionally caramel. The body shape of the Australian Tiffanie is similar to that of the Burmilla cat - with the difference that the Tiffanie is much more temperamental in comparison, has a wider head, a smaller body, and shorter legs. In addition, the ears and nose are smaller and the coat longer than in the Burmilla cat. The Australian Tiffanie was recognised by the Australian Waratah State cat Alliance in 1998.

Breeding start
1990s
Australia

ACF	ACFA	CAA	CCA-AFC	CCC	CCCA	CFA	FB	FIFe	GCCF	LOOF
✓							✓	✓	✓	✓

NZCF	SACC	TICA	WCF
✓	✓		✓

Selection of the most important cat associations in the world: Australia, Canada, China, France, Germany, New Zealand, South Africa, UK and USA

ATELIER KAYMAK®
Science
PAPERS

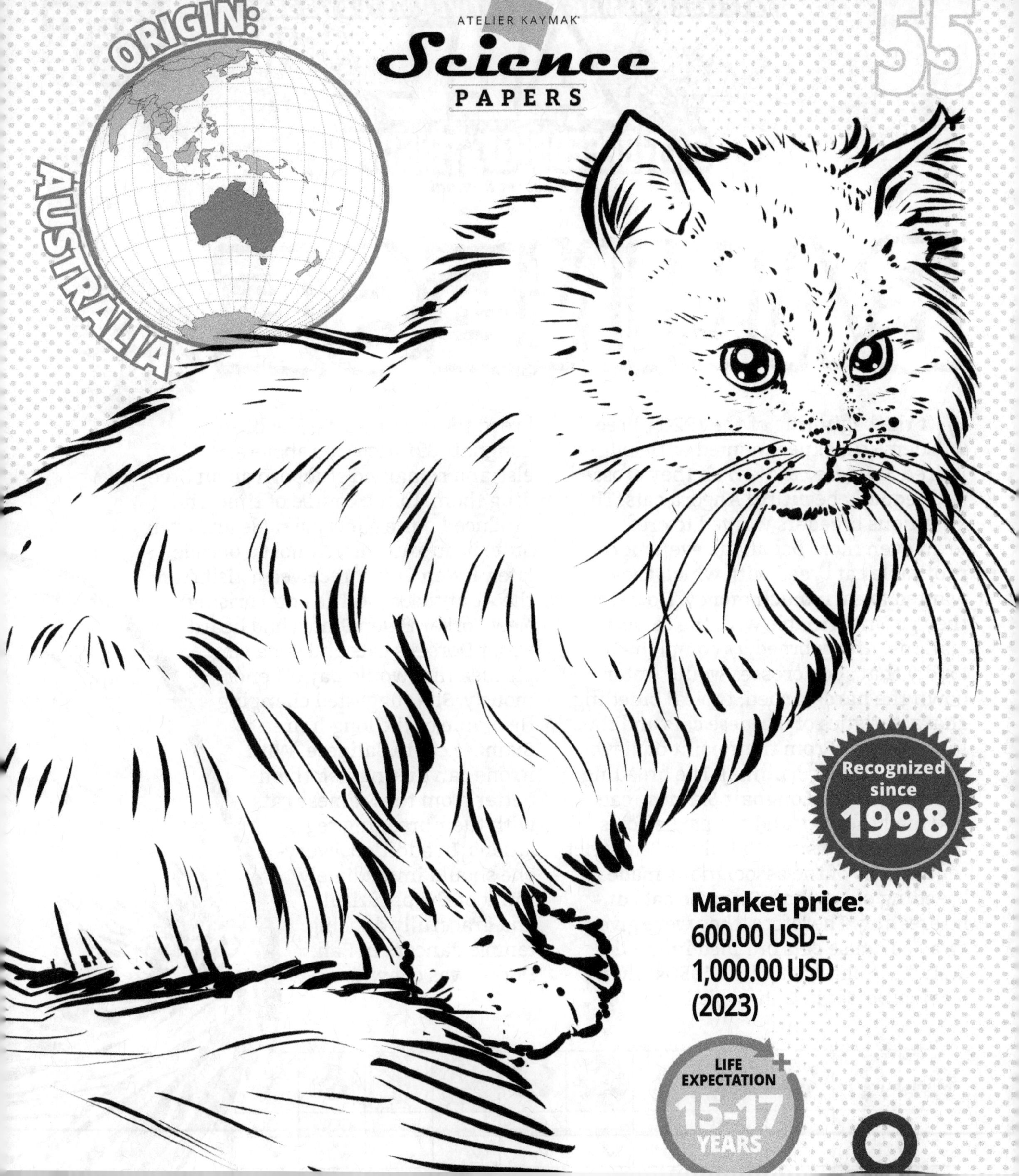

Recognized since 1998

Market price:
600.00 USD–
1,000.00 USD
(2023)

LIFE EXPECTATION
15-17
YEARS

Breeding start 1950 United States

BALINESE

Recognized since 1960

At the beginning of the 1920s, breeders still experimented with longhaired Siamese cats, which they crossed with equally beautiful Angora cats. The ambitious breeders wanted to create an even more beautiful, even more profitable cat breed with which they wanted to earn much money. However, they got into trouble with the cat associations. Angry breeders complained loudly that such crosses would water down the hard-earned, typical breeding characteristics of Siamese cats and thus falsify them. From their point of view, this was almost criminal. The breeding program of the Longhair Siamese cats had died thereby, and the past efforts were thereby a smooth failure. The strict regulations of the associations made the breeding of longhair Siamese cats uninteresting. Therefore, they were given away as house cats or killed immediately. This changed in the 1950s when Marion Dorset in California began to breed those same cats. She liked Longhair Siamese cats above all else and remained stubborn about breeding them. In the course of time, she produced quite a few passable and even quite beautiful cats. Almost a decade later, Helen Smith received a visit from the Atlantic coast. The enterprising New Yorker Helen Smith had heard about Dorset's cats and got a simple idea that would pay off enormously. She suggested changing the name of the long-haired Siamese cat to Balinese. With it, one can distinguish them better from the Siamese cats with their breeding regulations. In addition, everyone should immediately think of the beautiful and gracefully dancing temple dancers in Bali. Dorset was convinced.

ACF	ACFA	CAA	CCA-AFC	CCC	CCCA	CFA	FB	FIFe	GCCF	LOOF
✓	✓	✓	✓	✓	✓	✓	✓	✓	✓	✓

NZCF	SACC	TICA	WCF
✓	✓	✓	

Selection of the most important cat associations in the world: Australia, Canada, China, France, Germany, New Zealand, South Africa, UK and USA

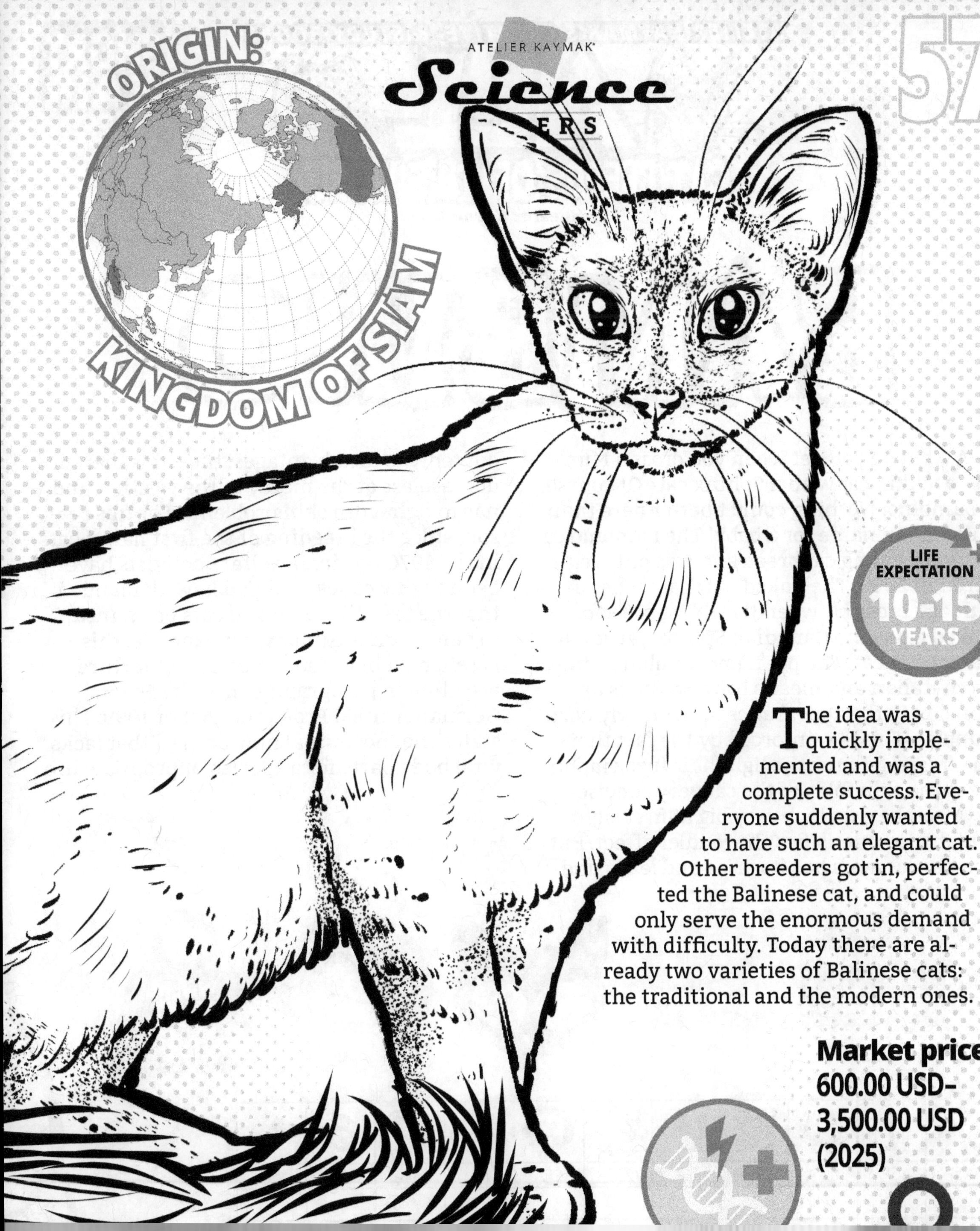

LIFE EXPECTATION
10-15
YEARS

The idea was quickly implemented and was a complete success. Everyone suddenly wanted to have such an elegant cat. Other breeders got in, perfected the Balinese cat, and could only serve the enormous demand with difficulty. Today there are already two varieties of Balinese cats: the traditional and the modern ones.

Market price:
600.00 USD–
3,500.00 USD
(2025)

BAMBINO

What do you get when you cross a Munchkin cat with a Sphinx cat? Of course: a bambino! Nothing could t better here than this Italian name for a baby! The Bambino cat is a very young cat breed and was only registered in 2005. The naked hair cat is the result of a cross between the oldest breed of all naked cats - the Canadian Sphinx cat and a Munchkin cat. The pet name Bambino comes from Italian and means baby, which is an allusion to the appearance of the newly born puppies. This is reinforced by the fact that the missing fur or the light hair fuzz emphasizes the skin folds of the cat very concisely. Also, the ears look very big at first sight, which also intensifies the toddler effect. But the hairlessness has its price and leads undoubtedly to disadvantages: the cat is thereby defenseless at the mercy of the weather and has to fight with skin problems. For this reason, since the breeding of the first naked cats in the 1970s, animal welfare activists have been highly concerned and have demanded that the breeding of naked cats be banned in general. In Germany, for example, this is reflected in the so-called ban of torture breeding in paragraph 11 b of the Federal German Animal Protection Act of 1999. This states that no animal may be bred that lacks vital body parts for a species-appropriate life.

Breeding start
2000s
United States

HAIRLESS CAT

LIFE EXPECTATION
9-15
YEARS

Weight
5-9
lbs

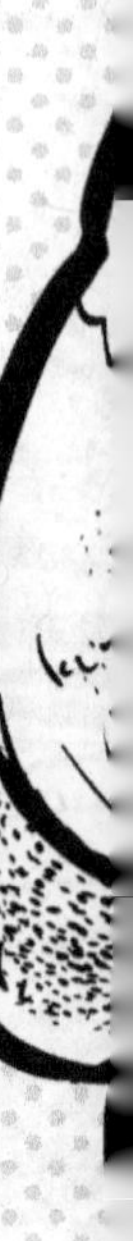

ACF	ACFA	CAA	CCA-AFC	CCC	CCCA	CFA	FB	FIFe	GCCF	LOOF

NZCF	SACC	TICA	WCF
		✓	

Selection of the most important cat associations in the world: Australia, Canada, China, France, Germany, New Zealand, South Africa, UK and USA

ATELIER KAYMAK
Science
PAPERS

Breeding start
1980s
United States

BENGAL

The Bengal cat carries its name because of its beautiful fur, which makes it look like a mini leopard. She is the Offspring of hybrid cats that come from crossing a female Asian leopard cat (Prionailurus bengalensis) with a black street cat. The Asian leopard cat is from the family of small cats and is thus from the genus small cats. Small cats should not be underestimated: they are small cats by name, but actual wild cats. So, predators are much closer to lynxes and pumas than domestic cats. Nevertheless, it is evident that humans have tried to mate both species just because of their size. Allegedly there were attempts to cross wild cats with domestic cats in 1889 and 1924. It is proven that in the 1963s, the Californian geneticist Jean Sugden mated a small Asian leopard cat with an equally small, tame domestic cat, a black tomcat. In 1972, she took over eight female hybrid cats from her colleague, the geneticist Dr. Willard Centerwall. The two worked together on research into the feline leukemia virus (FeLV) at the University of California. The Asian leopard cat is suitable for study because it has the same chromosome number as domestic cats. Although puppies of hybrid cats are usually born infertile, in this case, there were healthy female kittens in the litter.

ACF	ACFA	CAA	CCA-AFC	CCC	CCCA	CFA	FB	FIFe	GCCF	LOOF
✓	✓	✓	✓	✓	✓	✓	✓	✓	✓	✓

NZCF	SACC	TICA	WCF
✓	✓	✓	✓

Selection of the most important cat associations in the world: Australia, Canada, China, France, Germany, New Zealand, South Africa, UK and USA

ORIGIN:

SOUTHEAST ASIA, USA

According to breeding guidelines, at least four generations of puppies must be conceived before the Bengal cat breed standard is reached. The Bengal cat has either a rosette, striped, spotted, or marbled coat. She is very chatty and playful and has no fear of water, and loves it very much: she plays with the dripping tap whenever she can or jumps into the bathtub. Because the Bengal cat has a mad bounce, keeping it in an outdoor enclosure is highly recommended.
The Bengal cat was recognized by the TICA in 1983.

Market price:
200.00 USD–12,000.00 USD (2025)

HYBRID

TOP 10

Recognized since 2013

LIFE EXPECTATION 12-16 YEARS

DESIGNERCAT

LUXUS

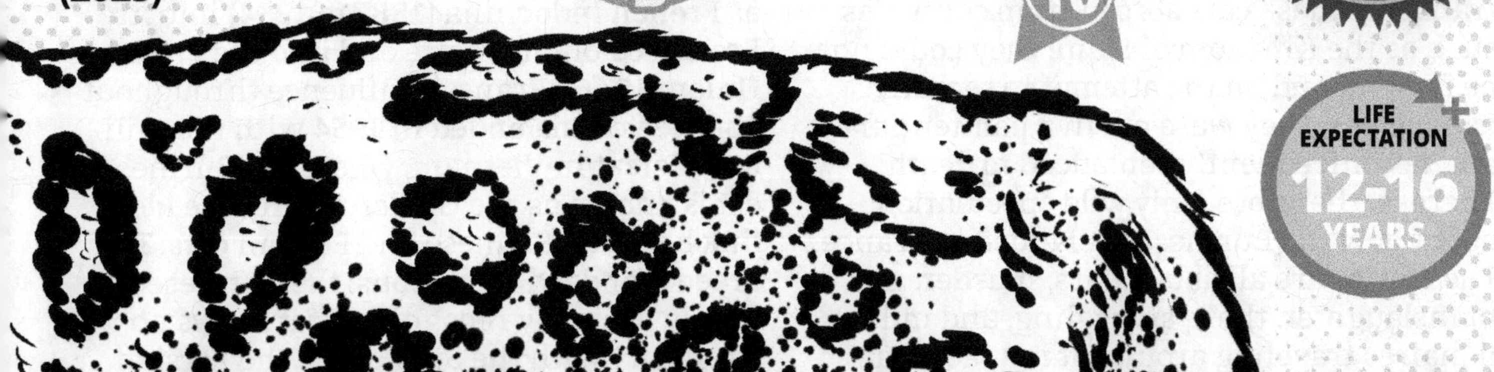

BIG

BURMESE CAT

Breeding start
1900s
France

The roots of the gentle Burmese cat - or Sacred Burma - lie in Southeast Asia's temples deep in the jungle. For a long time, it was practically impossible for foreigners to transport these cats abroad from countries such as the Kingdom of Siam: they could not be bought, and on the attempt to secretly export cats, they were put in a penitentiary and not infrequently sentenced to death. There is, therefore, only wild speculation about how the Burmese cat landed in France. The stories are all about wars, murder, and manslaughter, theft, smuggling, and millionaires traveling around. It is a fact that in the European history of cat breeding - Sacred Burma appeared for the first time in French stud books from 1915 - the breed was registered as an independent breed in 1925. Modern breeding of the Burmese cat began in France in 1955. French troops had just returned home from the Indochina War, which raged from 1946 to 1954 and led to the Vietnam War (1955 to 1975), almost three times as long and even more terrible. France - La Grande Nation - had annexed the home countries of the Burmese cat at the end of the 19th century: Annam, Cochinchina, Cambodia, the kingdom of the Khmer, Laos, and Vietnam - and incorporated them as French Indochina (1887 to 1954) into its French colonial empire and into the French Union (1946). France's influence throughout Southeast Asia ended in 1954 with the military defeat in Vietnam, where the Burmese cat is known as the Golden Mean - the ideal link between Siamese and Persian cats. The breed standard is rigorous: the Burmese cat is only officially recognized if it shows no similarity with the Siamese or the Persian cat. It belongs to the partial albinos of the so-called point cats. Therefore, it has very light, silky short-haired fur without an undercoat. She has radiant, deep blue eyes. Points cats - are so-called because of their coat, because the otherwise very light fur in the cooler regions - at the so-called points: nose, face, ears, and extremities - runs out in dark brown to black and looks like gloves and boots.

ACF	ACFA	CAA	CCA-AFC	CCC	CCCA	CFA	FB	FIFe	GCCF	LOOF
✓	✓	✓	✓	✓		✓			✓	

NZCF	SACC	TICA	WCF
✓	✓	✓	

Selection of the most important cat associations in the world: Australia, Canada, China, France, Germany, New Zealand, South Africa, UK and USA

ANCIENT ORIGIN

ATELIER KAYMAK
Science
PAPERS

Recognized since 1925

ORIGIN:

SOUTHEAST ASIA, FRANCE

LIFE EXPECTATION
14-16
YEARS

Market price:
150.00 USD–2,800.00 USD
(2025)

TOP 10

BOHEMIAN REX

The Bohemian Rex cat is the only pedigree cat that was developed in the Czech Republic - more precisely: in the predecessor state Czechoslovakia. At the beginning of the 1980s, a curly Rex cat was born in Liberec, Bohemia. She was conceived by the sister Persians Apolonia and Adonis, who became the founders of the Bohemian Rex breed. The Bohemian Rex cat owes its name to its origins in Bohemia, a region bordering (clockwise) Silesia in the northeast, historical Moravia in the east, Austria in the south, Bavaria in the southwest/west, and Saxony in the northwest. By the way, the diverse environment with the resulting linguistic confusion was the reason why the common expression Bohemian village or Bohemian village for I don't understand that has found its way into the German vocabulary. The Bohemian Rex cat owes the curly coat to the gene defect of the same name, which is noticeable from the mutation of the keratins by the gene krt71.

ACF	ACFA	CAA	CCA-AFC	CCC	CCCA	CFA	FB	FIFe	GCCF	LOOF
							✓	✓		

NZCF	SACC	TICA	WCF

Selection of the most important cat associations in the world: Australia, Canada, China, France, Germany, New Zealand, South Africa, UK and USA

As a result, the cat does not grow any undercoat worth mentioning - usually even no undercoat at all, which manifests itself as a tight-fitting, thin coat. The mutation is recessive and leads to the fact that Bohemian Rex Puppies are occasionally found in litters of Persian cats as well as in Maine Coons. For a long time, the Czech Bohemian Rex cat stood hidden in the shadow of its more famous cousins from West Europe: the Cornish Rex-, Devon Rex-, LaPerm- and Selkirk Rex- cat. But also, the California Rex, German Rex-, Ohio Rex-, Oregon Rex-, Ural Rex-, Tennessee Rex-, Tasman Rex- and Skookum cat. But in the meantime, the Bohemian Rex cat is known worldwide. FIFe added the Bohemian Rex cat to its register in 1994.

Recognized since **2013**

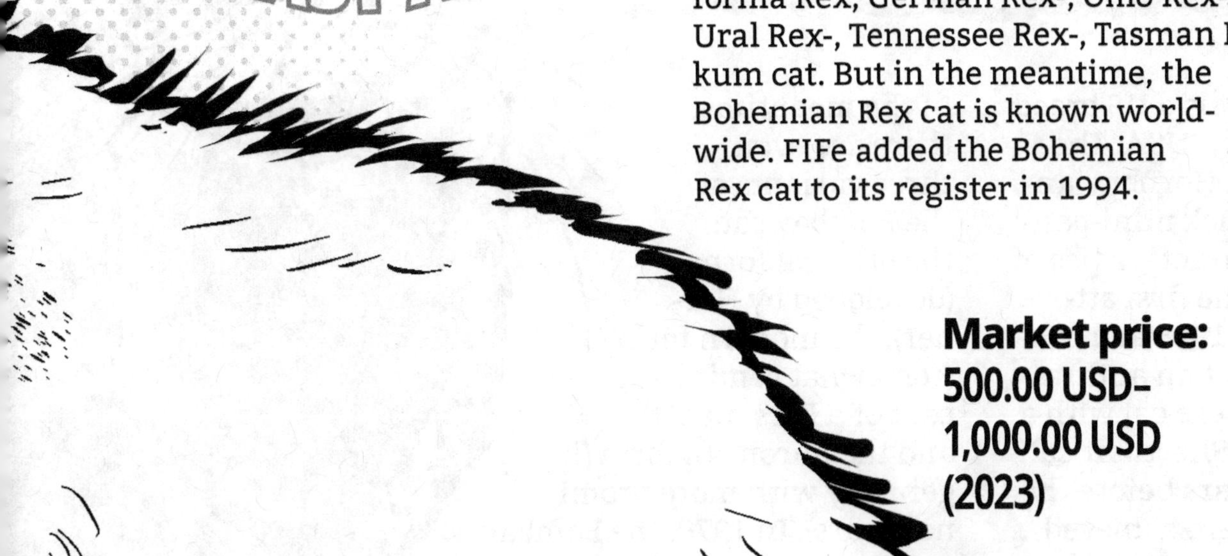

Market price:
500.00 USD–
1,000.00 USD
(2023)

Breeding start **1980s** United States

LIFE EXPECTATION **13-17** YEARS

SHORT HAIR

BOMBAY

The Bombay cat is a US breed from Kentucky/USA. The cat breeder Nicki Horner wanted to create a black mini-panther with the characteristics of a domestic cat. The first attempts started in 1958 and consisted of crosses of an anthracite-brown Burmese cat with a black, American Shorthair cat. It took several years before the first successes were achieved in 1965, and Horner was able to present his cats to an astonished audience. A few years later, there were efforts by breeders in Great Britain to cross European Burmese cats with black British Shorthair cats. However, this led to variations, which in turn led to controversy on both sides of the Atlantic. In the meantime, there are even three variants of the Bombay cat: the original form (developed by Horner), the modern form (somewhat reminiscent of a Persian cat), and the European form (larger body with more prominent ears). In 1970, the Bombay cat was recognized by the CFA. The small, handsome Bombay cat bears the short black coat of the American Shorthair cat and the copper eyes of the Burmese cat. Because of its fabulous gait, it has been nicknamed the Mini Panther.

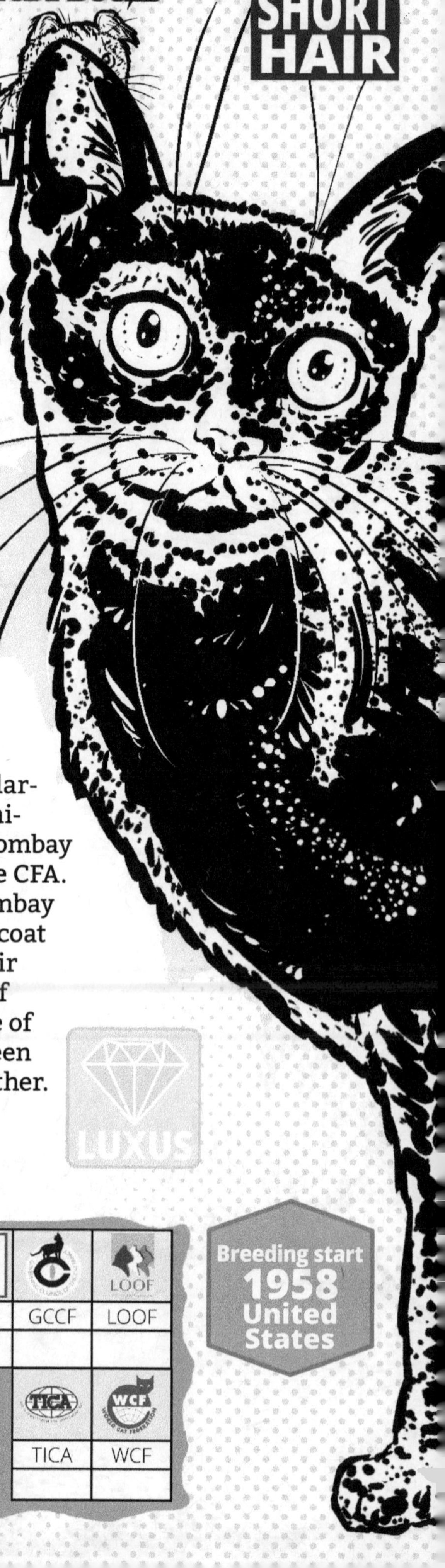

LUXUS

ACF	ACFA	CAA	CCA-AFC	CCC	CCCA	CFA	FB	FIFe	GCCF	LOOF
							✓	✓		

NZCF	SACC	TICA	WCF

Selection of the most important cat associations in the world: Australia, Canada, China, France, Germany, New Zealand, South Africa, UK and USA

Breeding start 1958 United States

Although the Bombay cat is much smaller than a standard domestic cat, it confidently expresses its heritage in its gait: The deep-shouldered pace is just like that of a full-grown panther. The lithe and muscular cat is very rare outside the USA. It is considered to be an amiable and affectionate creature that people would like to keep as a family pet.

Market price:
600.00 USD–2,500.00 USD (2025)

Recognized since 1970

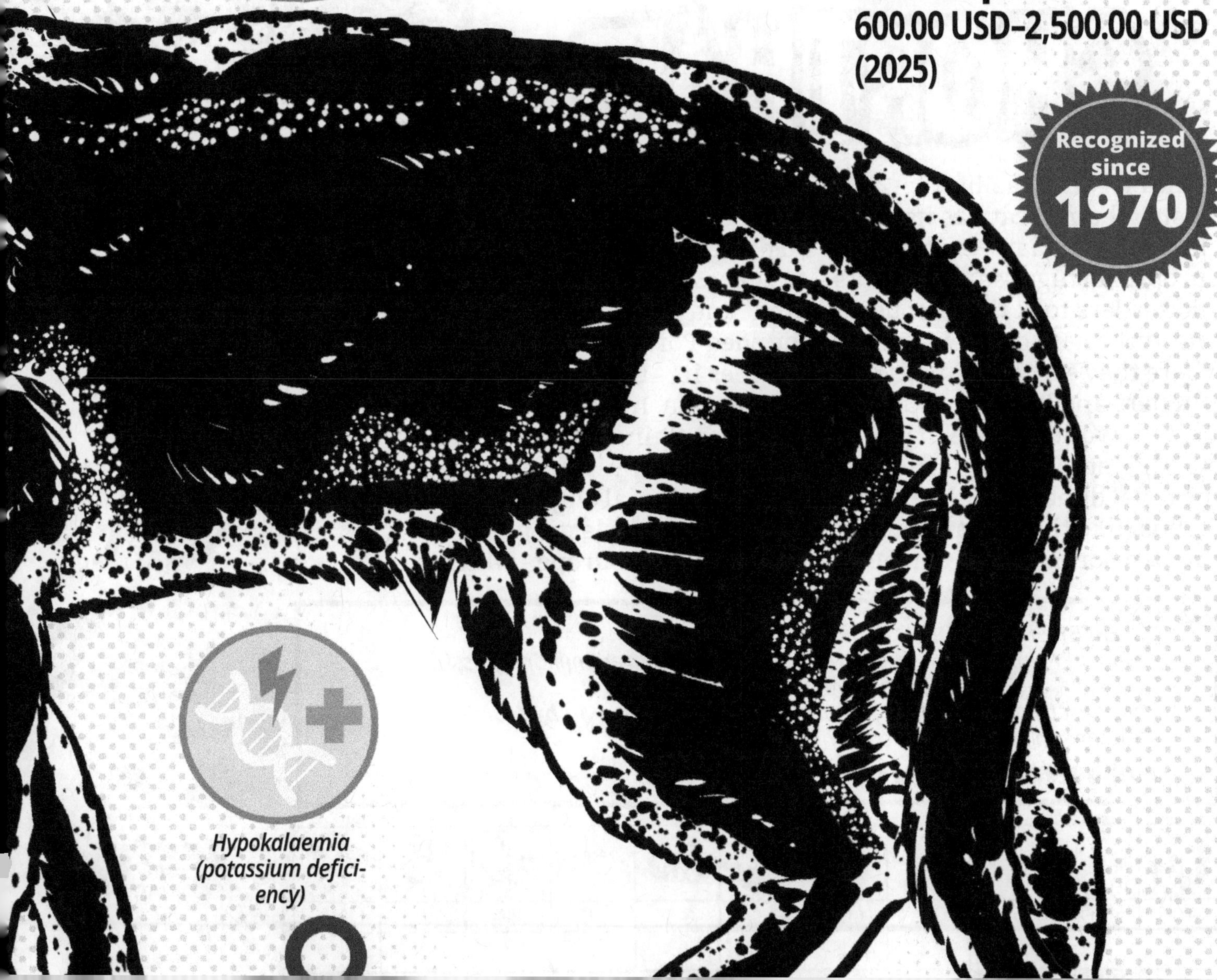

Hypokalaemia (potassium deficiency)

ATELIER KAYMAK'S CREATIVE- & STUDY BOOKS

ALL CAT BREEDS OF THIS W...

by Nuesret Kaymak

SHORT HAIR

BRAZILIAN SHORTHAIR

The elegant Brazilian shorthaired cat has its origin on the Iberian Peninsula in Europe. Portuguese conquerors and settlers took their robust cats with them on their ships, where they served as mouse Catchers and helped to build the first world empire in human history. It all began when the royal dynasty of the House of Avis (like the US car rental company of the same name) sent out their scout Captain Vasco da Gama and took possession of land* and gradually founded Portuguese colonies in America, Africa, Arabia, India, Southeast Asia, and China. In the course of its 500-year history, the Portuguese colonial empire formed the longest-lasting colonial empire in Europe - which

* *"I HEREBY CLAIM INDIA FOR BRITAIN!"*
"You can't claim us. We live here! There's five hundred million of us."
"Oh. I see. Well...Do you have a FLAG?"
"We don't need a bloody flag, this is our country, you b#stard!"
"No flag, no country! You can't have one."
Eddie Izzard, Dress to Kill

Breeding start
1980s
Brazil

LIFE EXPECTATION
14-20
YEARS

ACF	ACFA	CAA	CCA-AFC	CCC	CCCA	CFA	FB	FIFe	GCCF	LOOF

NZCF	SACC	TICA	WCF
			✓

Selection of the most important cat associations in the world: Australia, Canada, China, France, Germany, New Zealand, South Africa, UK and USA

began in 1415 with the conquest of the city of Ceuta in North Africa and only came to an end at the turn of the millennium with the abandonment of the last colony, the gambling island of Macau - also known as the Las Vegas of the East and the handover to China in 1999. The cats did their jobs all the time during the journey overseas and also went ashore with the people in the New World, in Brazil. For centuries, the Offspring of these cats could reproduce relatively undisturbed and lived as ordinary street cats. Then, in the 1980s, the cat breeder Dr. Paulo Samuel Ruschi became aware of them and started an official breeding program with them. In 1998, the Brazilian Shorthair cat was finally recognized as its own cat breed.

Market price:
500.00 USD–
1,400.00 USD
(2023)

Recognized since 1998

BIG CAT

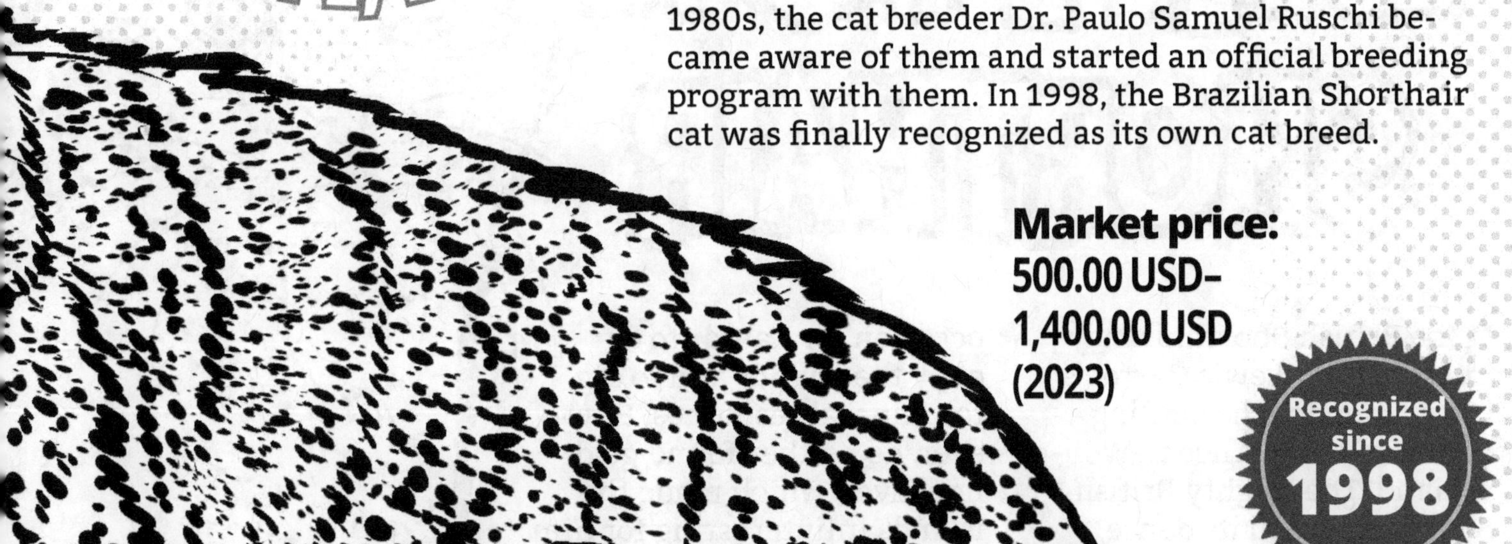

BRITISH SHORTHAIR

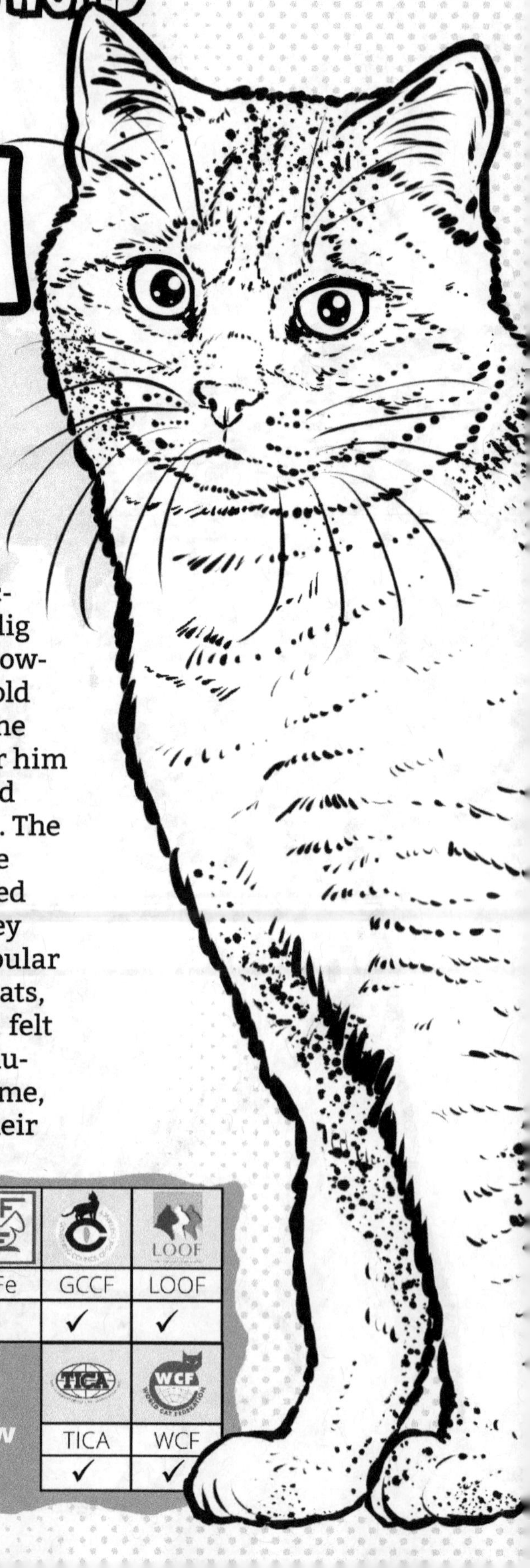

The British Shorthair may have been Lewis Carroll's inspiration for the smiling Cheshire cat in Alice in Wonderland. The mighty British Shorthair - with its dense, short coat - is one of the oldest cat breeds. In the first century, Roman invaders brought their Egyptian cats with them to the islands during Caesar's Britain campaigns in the Gallic War. Belgian chieftains fled to the Britons, the Island Celts, and asked them to ally themselves with them. This was a welcome occasion for Caesar to occupy the islands and to dig for mineral resources. However, he neither found gold nor silver, which made the islands uninteresting for him and drastically shortened the period of occupation. The cats brought along by the occupiers, however, mixed with the native ones. They quickly became very popular because the ponderous cats, with their gentle nature, felt quite comfortable near humans. In the course of time, they formed cats with their

ACF	ACFA	CAA	CCA-AFC	CCC	CCCA	CFA	FB	FIFe	GCCF	LOOF
✓	✓	✓	✓	✓	✓	✓	✓	✓	✓	✓

NZCF	SACC	TICA	WCF
✓	✓	✓	✓

Selection of the most important cat associations in the world: Australia, Canada, China, France, Germany, New Zealand, South Africa, UK and USA

own characteristics. In the 19th century, England began to breed and optimize exactly these cats with success: the specimens exhibited in a cat show in 1871 attracted worldwide attention. The British Shorthair cat has since been the most popular breed of cat in the British Isles. The British Shorthair cat is broadly built and has a large head with round eyes. Almost every color variation can be found, but the most iconic is probably the British blue.

Market price:
200.00 USD–7,000.00 USD (2025)

Recognized since 2009

Breeding start 1800s United Kingdom

ANCIENT ORIGIN

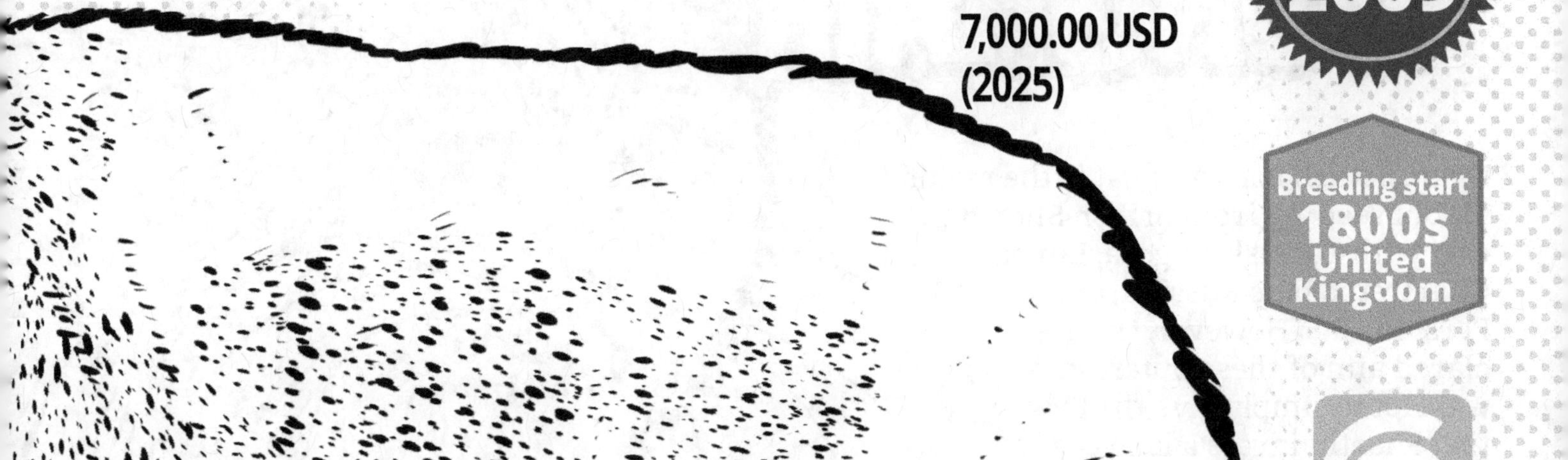

(with pure white coat): congenital hearing loss, deafness,

LIFE EXPECTATION 14-16 YEARS

BRITISH LONGHAIR

The British Longhair cat is the result of breeding from British Shorthair cats with Turkish Angora, Burma, Chartreux, blue Russian, and Persian cats. However, the declared aim of these outcrosses was not to simply give the British Shorthair cats a long coat. No, the breeders wanted to provide the roundish head of the British Longhair an elegant touch and add the flatter facial features of the Persian.

Breeding start 1910s United Kingdom

Recognized since 2009

LIFE EXPECTATION 12-14 YEARS

ACF	ACFA	CAA	CCA-AFC	CCC	CCCA	CFA	FB	FIFe	GCCF	LOOF
	✓	✓	✓				✓	✓	✓	✓

NZCF	SACC	TICA	WCF
	✓	✓	

Selection of the most important cat associations in the world: Australia, Canada, China, France, Germany, New Zealand, South Africa, UK and USA

ATELIER KAYMAK
Science
PAPERS

Since the Second World War, however, British Longhair cats were often the unloved black sheep of the English breeding programs. Only the recognition as a new cat breed by FIFe and Felis Britannica brought the break.

LUXUS

Autosomal dominant polycystic kidney disease (ADPKD)

Market price:
200.00 USD–7,000.00 USD (2025)

BURMILLA SHORTHAIR

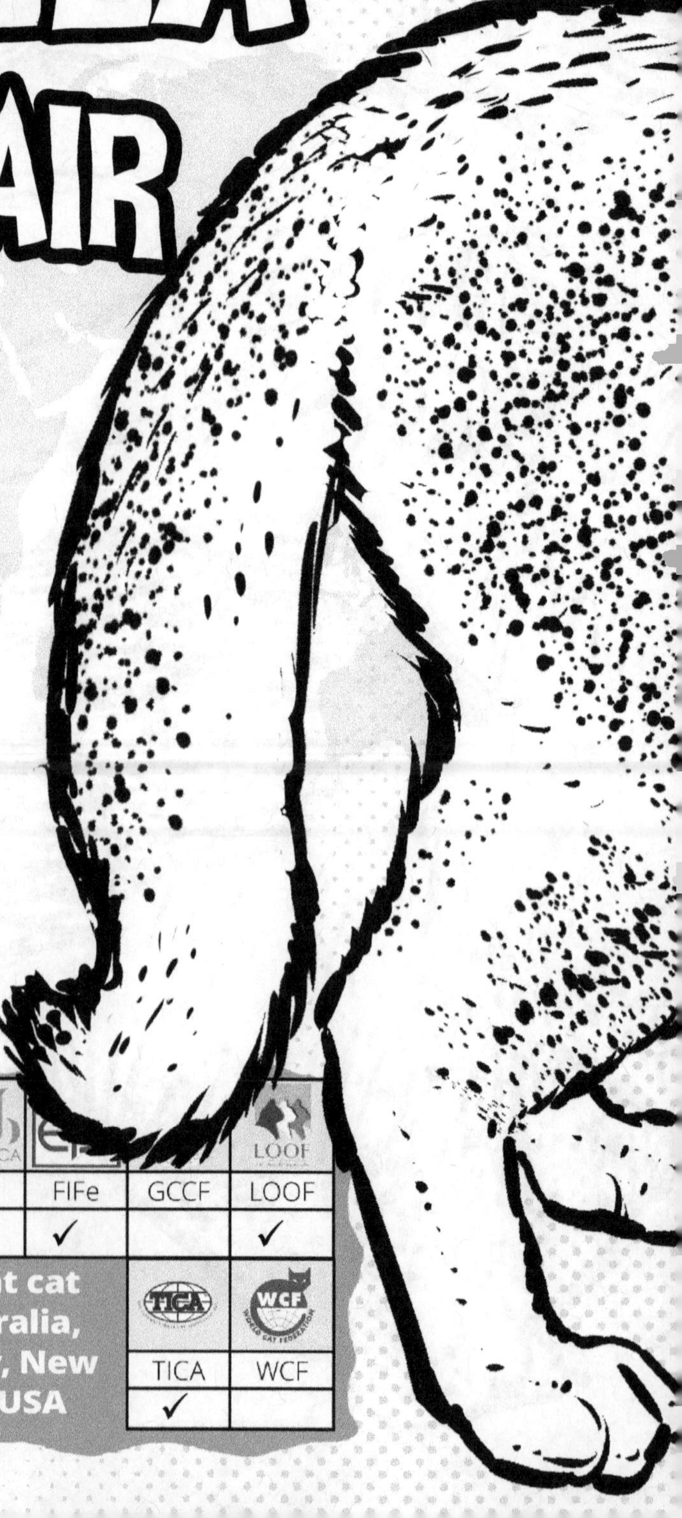

The Burmilla cat is very similar to the Burmese cat with its dense and exquisite soft silver fur. The Burmilla (a portmanteau of Burmese cat and Chinchilla cat) was founded by Mrs. Miranda Bickford-Smith in 1981 from the accidental crossing of a Persian chinchilla male named Jemari Sanquist and the Burmese cat named Bambino Lilac Fabergé. Another breeder, Thérèse Clark, acquired a puppy from the original litter and concentrated on the development of the breed coat, while Bickford-Smith focused on the development of the body shape. The breed received championship status in Great Britain in 1990 and is recognized as an experimental breed in Great Britain. While it is still rare in America, TICA recognized it in 2015.

ACF	ACFA	CAA	CCA-AFC	CCC	CCCA	CFA	FB	FIFe	GCCF	LOOF
✓	✓	✓	✓	✓	✓	✓	✓	✓		✓

NZCF	SACC	TICA	WCF
✓		✓	

Selection of the most important cat associations in the world: Australia, Canada, China, France, Germany, New Zealand, South Africa, UK and USA

ORIGIN: SOUTH ASIA, UK

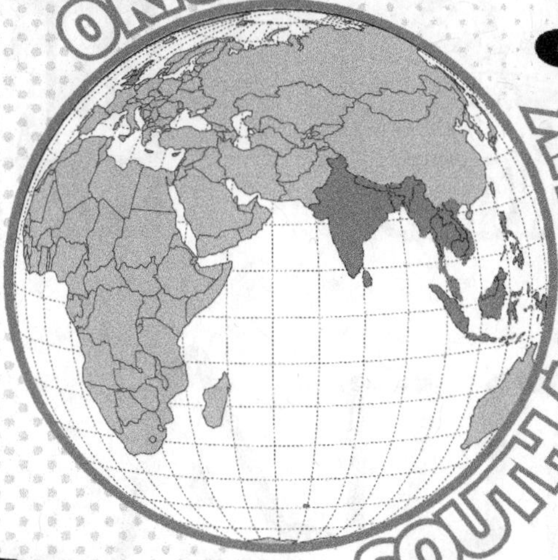

Breeding start
1981
United Kingdom

Recognized since 1994

Market price:
100.00 USD–
1,200.00 USD
(2023)

LIFE EXPECTATION

BURMILLA LONGHAIR

Have you ever seen a cat with more beautiful makeup? But the spirited longhaired Burmilla cat is not only famous because of her graceful beauty and elegance - but also because of her intelligence: this cat is so intelligent that she effortlessly plays poker with you when asked nicely. Her half-long coat follows the lines of her body and gives her a magnificent appearance. With her radiant green eyes and her friendly nature, the Burmilla cat wins the hearts of her owners uncompromisingly. She is affectionate: sometimes she has the habit of waiting at the door for her holder - and of demanding road toll in the form of cuddling units. Because the hind legs are slightly longer than the front legs, she looks as if she is jumping on your arms at any moment. The Burmilla cat never tires of demanding proofs of love. She is incredibly sensitive and intuitively notices how her owner is doing. Then the delightful chunk cheers up his owner with jokes and distracts him with acrobatic tricks.

Breeding start 1981 United Kingdom

LIFE EXPECTATION 14-16 YEARS

Weight 6-13 lbs

ACF	ACFA	CAA	CCA-AFC	CCC	CCCA	CFA	FB	FIFe	GCCF	LOO
✓	✓	✓	✓	✓	✓	✓	✓	✓		✓

NZCF	SACC	TICA	WCF
✓		✓	

Selection of the most important cat associations in the world: Australia, Canada, China, France, Germany, New Zealand, South Africa, UK and USA

ATELIER KAYMAK
Science PAPERS

Recognized since 1994

Hypokalemia (potassium deficiency)

Market price:
100.00 USD–
1,200.00 USD
(2023)

CASHMERE

Long-haired - or more precisely: semi-long-haired Cashmere Bengal cat, exists as long as its short-haired cousin. There are cat associations in which the Cashmere Bengal cat is listed as a so-called Pardino. Just like its short-haired version, the Cashmere Bengal cat is highly active and playful. It can be trained like a dog to fetch certain things. She is brilliant, so you can even teach her tricks. And just like a dog, it is easy to walk on a leash. Who needs a dog? There is, however, a drop of bitterness. At the latest, after entry into sexual maturity, when the cat is fully grown - it becomes critical. Then it becomes increasingly difficult to master and control the natural shoots of the Bengal cat. If, for example, their hunting instinct breaks out, there is hardly any stopping. The animal goes through, and as a keeper, you can easily be dragged along in the air like a paper kite. Therefore, there are strict animal protection laws, which should guarantee that the Bengal cats are enough free space given around their urges species-appropriately to be able to live out. And to it belongs, for example, a sufficiently large free enclosure. The Bengal cat was approved for breeding in 1985, and the New Zealand Cat Fanciers Association classified it as a forerunner of a cat breed in 2013. TICA awarded it the championship status in 2017.

Breeding start
1980s
United States

ACF	ACFA	CAA	CCA-AFC	CCC	CCCA	CFA	FB	FIFe	GCCF	LO

NZCF	SACC	TICA	WCF
✓		✓	

Selection of the most important cat associations in the world: Australia, Canada, China, France, Germany, New Zealand, South Africa, UK and USA

ATELIER KAYMAK
Science
PAPERS

ORIGIN: SOUTH ASIA, UNITED STATES

Market price:
200.00 USD–
12,000.00 USD
(2025)

Recognized since 2013

LIFE EXPECTATION
12-16
YEARS

Flat-chested Kitten syndrom (FCKS), Progressive Retinal atrophy (PTR), Diseases of eyes, heart and joints

CALIFORNIA SPANGLED

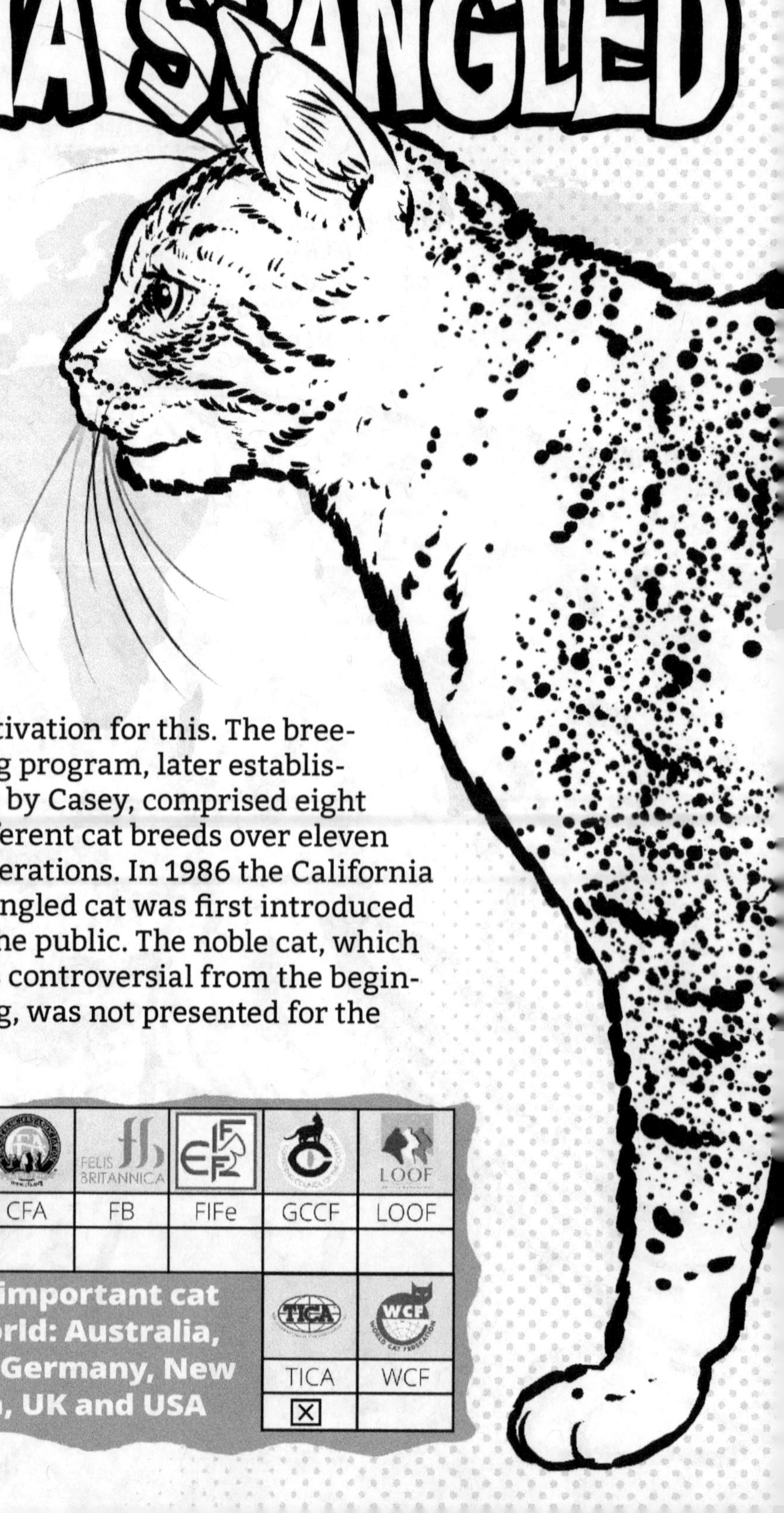

The beginning of the breeding of the California Spangled cat could have come from a film script: the screenwriter Paul Arnold Casey and his friend, the anthropologist Louis Leake, experienced the cruel death of a leopard by poachers in Tanzania in the 1970s. The experience was so traumatic that Casey and Leake decided to do something about it after returning to the United States. They decided to create a new domestic cat breed with the same coat pattern as the African cousins. Her intention was to change people's attitudes towards wearing exotic furs. They thought that no one would have the frivolity to wear their beloved domestic cat as a piece of clothing. So, fur coats should always be associated with beloved pets. Small, sociable house cats ocelots and house leopards should provide the necessary motivation for this. The breeding program, later established by Casey, comprised eight different cat breeds over eleven generations. In 1986 the California Spangled cat was first introduced to the public. The noble cat, which was controversial from the beginning, was not presented for the

ACF	ACFA	CAA	CCA-AFC	CCC	CCCA	CFA	FB	FIFe	GCCF	LOOF

NZCF	SACC	TICA	WCF
		☒	

Selection of the most important cat associations in the world: Australia, Canada, China, France, Germany, New Zealand, South Africa, UK and USA

first time at a cat show but appeared on the front page of a Christmas Catalogue of the luxury department store chain Neiman Marcus. It was also contentious because the cat was advertised in the Catalogue as a gift item. In 1991 the breed was recognized by the TICA. The relatively rare California Spangled cat (its creator, Casey, died in 2007 and thus also the official breeding program) is overshadowed by other exotic breeds such as the Ocicat cat and the Bengal cat. TICA, therefore, does not keep them in the register.

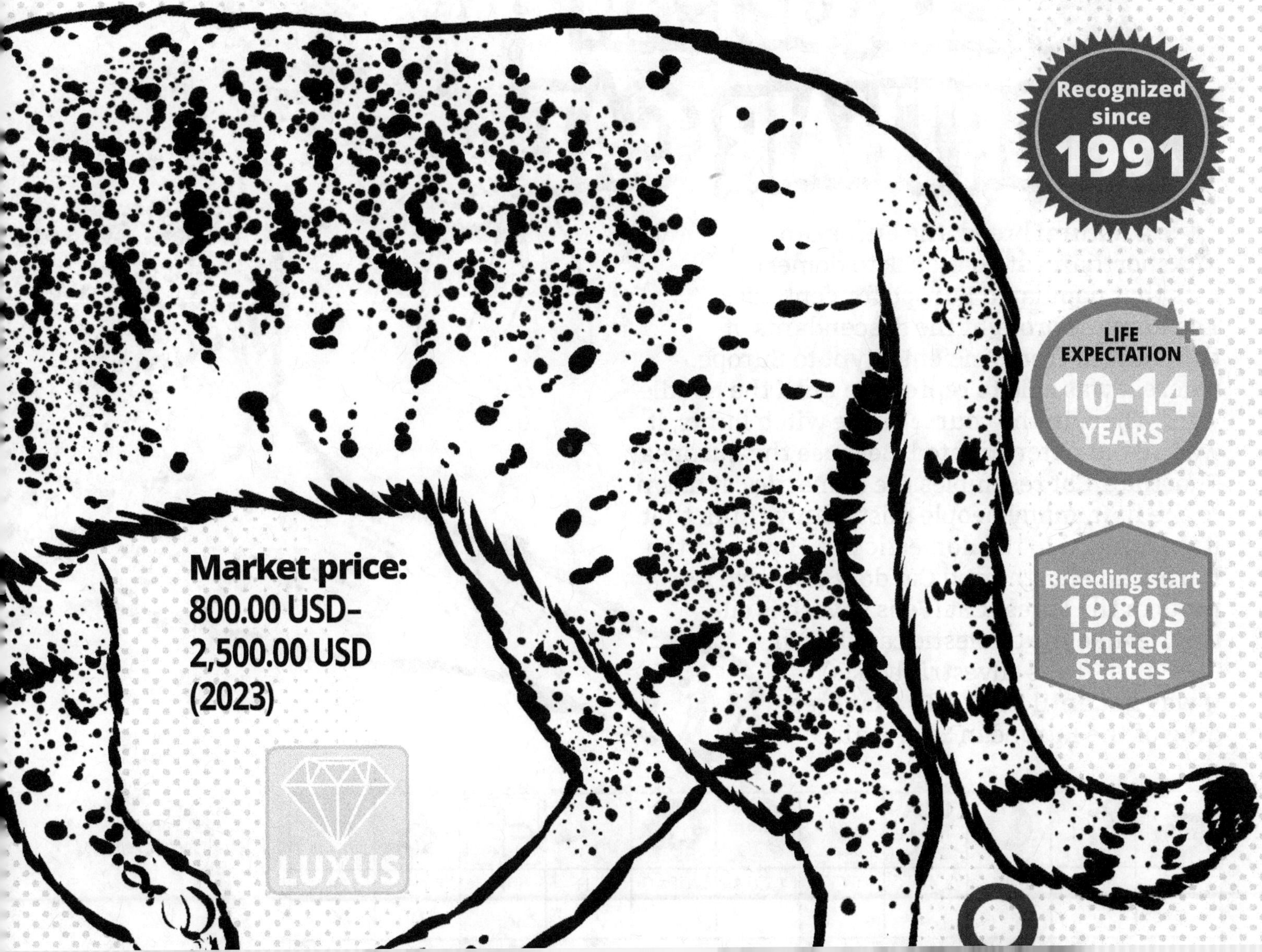

Recognized since 1991

LIFE EXPECTATION 10-14 YEARS

Breeding start 1980s United States

Market price:
800.00 USD–
2,500.00 USD
(2023)

LUXUS

SHORT HAIR

Breeding start **1950s** United States

CELTIC

ALSO EUROPEAN SHORTHAIR CAT

As a natural breed, the European shorthair cat goes back to domestic cats that populated Europe for centuries. The Romans brought the descendants of the wild cat from Ancient Egypt to Europe, where it was able to reproduce until the Middle Ages. Then, in the course of the witch hunts, it was all but exterminated. Because the European Shorthair Cat resembles the domestic cat down to the hair, many people mistakenly think that it is identical to the domestic cat. However, the European Shorthaired Cat does not have its roots in the Forest Cat (Felis silvestris silvestris) like the domestic cat - but in the Falcat (Felis silvestris lybica). It was not until the 1990s that breeding of the European Shorthair

ACF	ACFA	CAA	CCA-AFC	CCC	CCCA	CFA	FB	FIFe	GCCF	LOOF
							✓	✓		✓

NZCF	SACC	TICA	WCF
			✓

Selection of the most important cat associations in the world: Australia, Canada, China, France, Germany, New Zealand, South Africa, UK and USA

LIFE EXPECTATION 14-16 YEARS

Cat began again in Scandinavia, which has now grown to almost 100 varieties. The European Shorthair is not as common as the English and American Shorthair. In Germany the European Shorthair Cat is hardly known. In 1949, the French FIFe accepted the European Shorthair Cat as a separate breed, but there are indications that it was already recognised a decade earlier.

Recognized since 1882

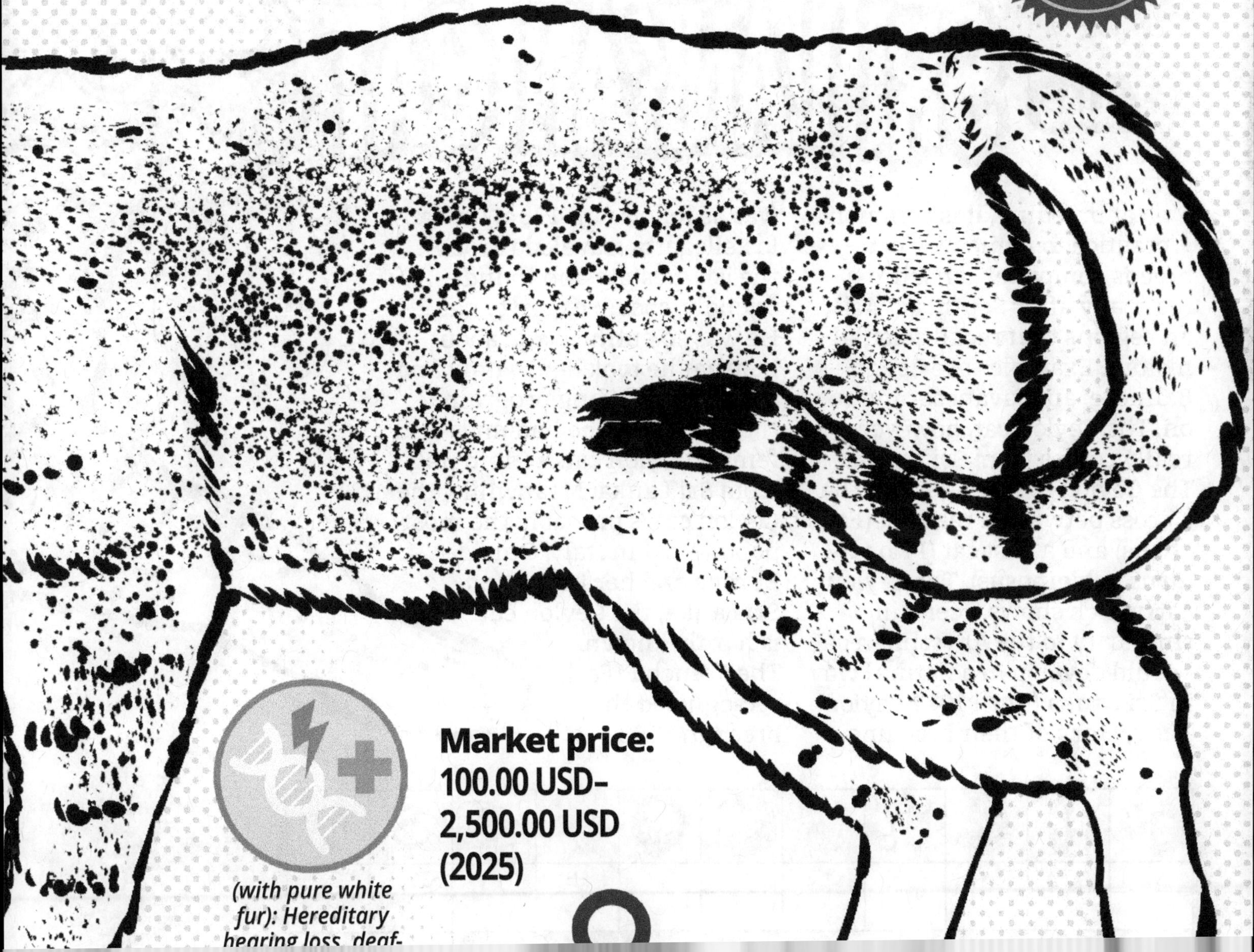

Market price:
100.00 USD–
2,500.00 USD
(2025)

(with pure white fur): Hereditary hearing loss, deaf-

CEYLON

ALIAS SRI LANKAN CAT

In cat breeding, it is common practice to name new cat breeds completely anachronistically after Southeast Asian regions that recall colonial history. Examples are Bengali, Balinese, and Javanese, and so on. The Ceylon cat, however, really comes from Sri Lanka. The Ceylon cat is believed to be a cross between a reed cat (Felis chaus) and a rust cat (Prionailurus rubiginosus). The Ceylon cat, which shows a certain similarity to the Abyssinian cat - could develop in a natural way in Sri Lanka (until 1972 Ceylon). The history of their breeding was as follows: In 1964, Dr. Paolo Pellegatta traveled through the island state and found a friendly, sociable, pretty cat. He was so enthusiastic about her that he took her to Italy and started a breeding program there. For this, he had male and female Ceylon cats sent to Italy from Sri Lanka. Meanwhile, the Ceylon cat is probably the most popular cat in Italy. Outside Europe and her homeland Sri Lanka, the Ceylon cat is hardly known. The French FIFe rec-ognized the breed in 1993.

ACF	ACFA	CAA	CCA-AFC	CCC	CCCA	CFA	FB	FIFe	GCCF	LOOF
										✓

NZCF	SACC	TICA	WCF
			✓

Selection of the most important cat associations in the world: Australia, Canada, China, France, Germany, New Zealand, South Africa, UK and USA

Breeding start
1984
Italie

ATELIER KAYMAK
Science
PAPERS
ORIGIN:
SRI LANKA, ITALIE
LIFE EXPECTATION
12-15
YEARS
Recognized since
1993
Market price:
1,000.00 USD–
2,500.00 USD
(2023)

CHARTREUX

The Chartreux cat is called Carthusian or Maltese cat in some countries. The name irritates because the cat has nothing to do with Malta, the monks, or their stay on the island. In truth, the pretty cat with the attentive amber eyes comes from the barren Rocky Mountains of Mesopotamia, an area that stretches over today's countries Iran, Syria, and Turkey. For a long time, the dense, beautiful fur of these muscular cats was processed into coats by the inhabitants, and their flesh was eaten, one suspects. Merchants who traveled to the Orient brought the elegant animal to France. One of the first more detailed descriptions of this cat can be found in the 1558 poem Vers Français sur la mort d'un petit chat (French verse about the death of a tiny kitten) by Joachim du Bellay. The Chartreux cat called smiling cat did not get its name until 1750 in the Chat des Chartreaux. Especially on the Breton island of Belle-Île-en-Mer there was a manageable, stable population of these pretty cats. In the 1920s, cat breeders started a breeding program with a small number of animals there. During the Second World War, only a

Breeding start
1700s
France

ACF	ACFA	CAA	CCA-AFC	CCC	CCCA	CFA	FB	FIFe	GCCF	LOOF
	✓	✓	✓	✓		✓	✓	✓	✓	✓

NZCF	SACC	TICA	WCF
		✓	✓

Selection of the most important cat associations in the world: Australia, Canada, China, France, Germany, New Zealand, South Africa, UK and USA

few specimens were found in France due to hunger and devastation. Therefore, in order to preserve the population, the British Shorthair cat was crossed in. The Chartreux is strongly built and has dense, soft fur. It has short limbs and copper eyes. The affectionate cat is not afraid of people and is known for being much attached to people and following them from room to room. The French President Charles De Gaulle once had a world-famous Chartreux, Gris-gris - which followed him on foot and kick.

Recognized since 1882

Market price:
200.00 USD–3,000.00 USD (2025)

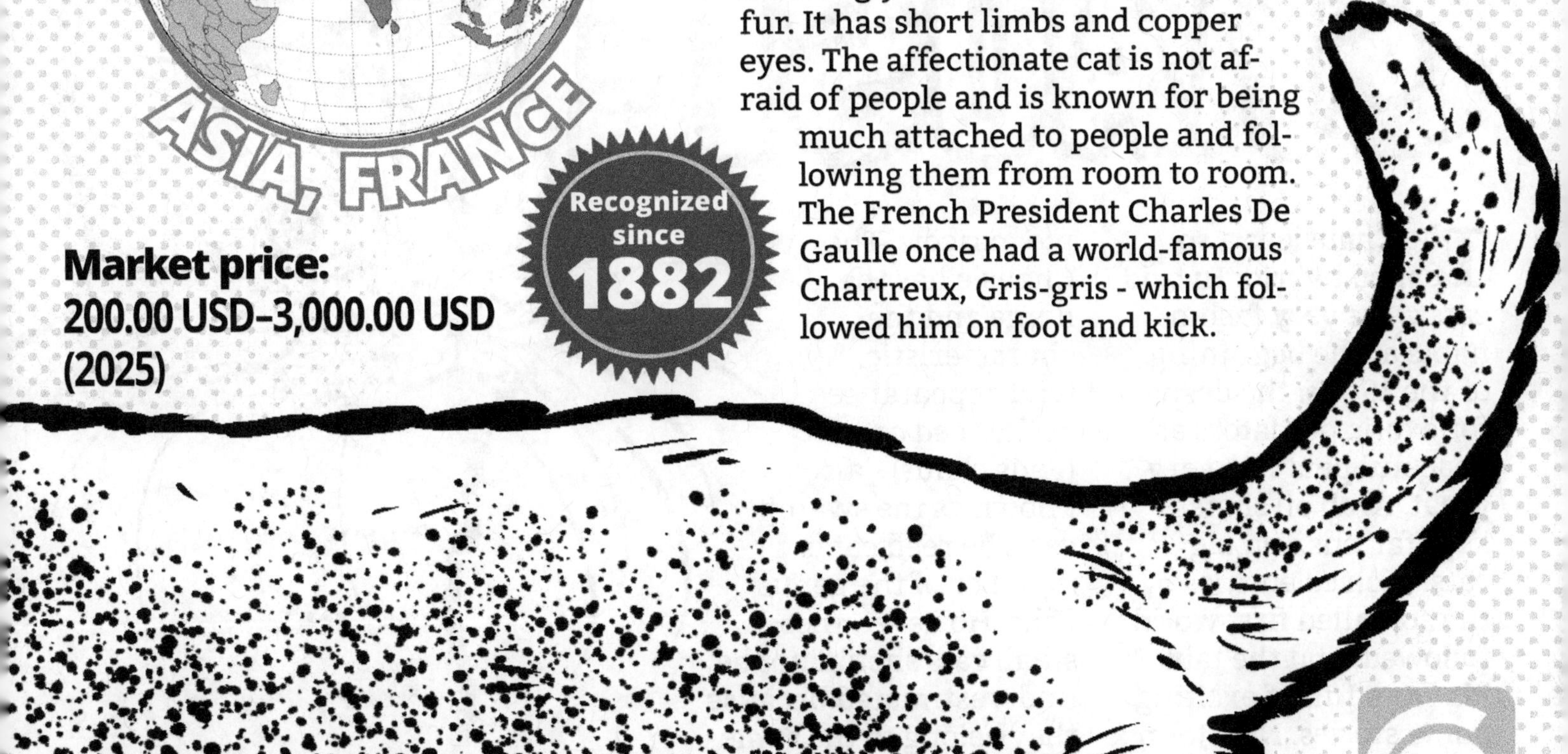

ANCIENT ORIGIN

LIFE EXPECTATION

CHAUSIE

The Chausie is a so-called hybrid. Hybrid cats were very fashionable at the beginning of the 1960s: the desire for exotic predators as peaceful pets was very great. Arab oil sheiks kept falcons and big cats. Rich celebrities of the so-called first world followed, and the tabloids were full of exciting home stories. Now John and Jane wanted to have such a tiger. Indeed, the many feature films like Hatari and TV series like Daktari also contributed to this, which increased the popularity of the wild cats: Big cats like Clarence - the cross-eyed lion from Daktari - became mascots. The Chausie has the name and the characteristic wild appearance of the reed cat (Felis chaus) - also known as the swamp cat. The reed cat is a member of the genus Felis. However, these small cats should not be underestimated because they belong to the same family as the predator cheetah, lynx, and cougar. The Chausie, on the other hand, is lovable and friendly. It lives in large parts of the Middle East, India, and Southeast Asia.

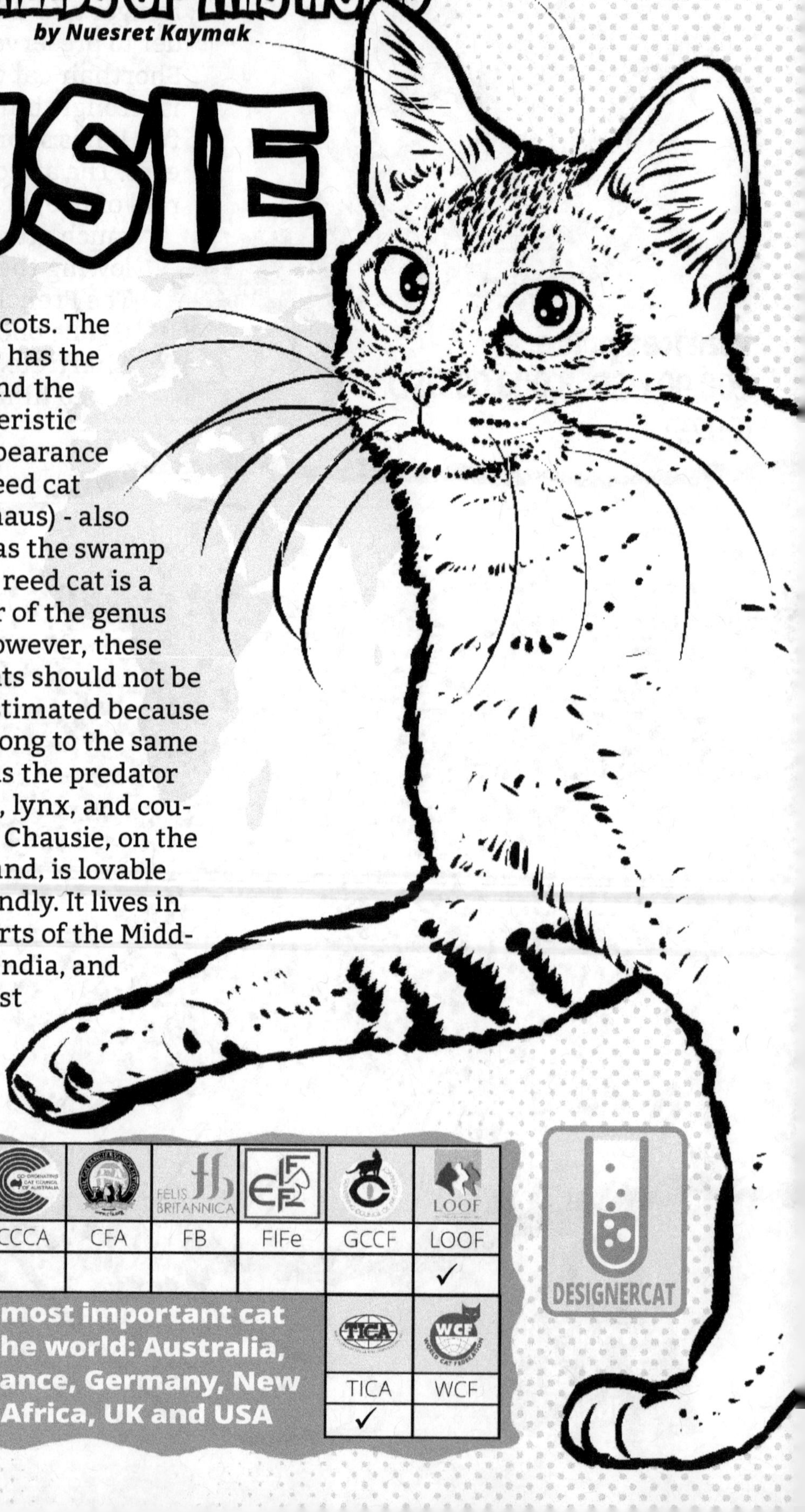

ACF	ACFA	CAA	CCA-AFC	CCC	CCCA	CFA	FB	FIFe	GCCF	LOOF
										✓

NZCF	SACC	TICA	WCF
		✓	

Selection of the most important cat associations in the world: Australia, Canada, China, France, Germany, New Zealand, South Africa, UK and USA

DESIGNERCAT

There are finds along the Nile delta that prove the existence of reed cats in ancient Egypt. From samples of mummified cats, one can conclude that reed cats lived together with domestic cats in ancient Egypt and were probably crossed together for thousands of years. The first breeding program of the present time took place in the 1960s. The Chausie was awarded championship status by TICA in 2013, with the CFA beginning the first phases of recognition. With high cheekbones and a long, elegant body, the Chausie resembles the Abyssinian cat somewhat. Their short coat can be offered in many variations, but in the world official competition only solid black coat and black and brown pattern is accepted.

Breeding start
1960s
France

BIG CAT

Recognized since 2013

HYBRID

LIFE EXPECTATION
12-14
YEARS

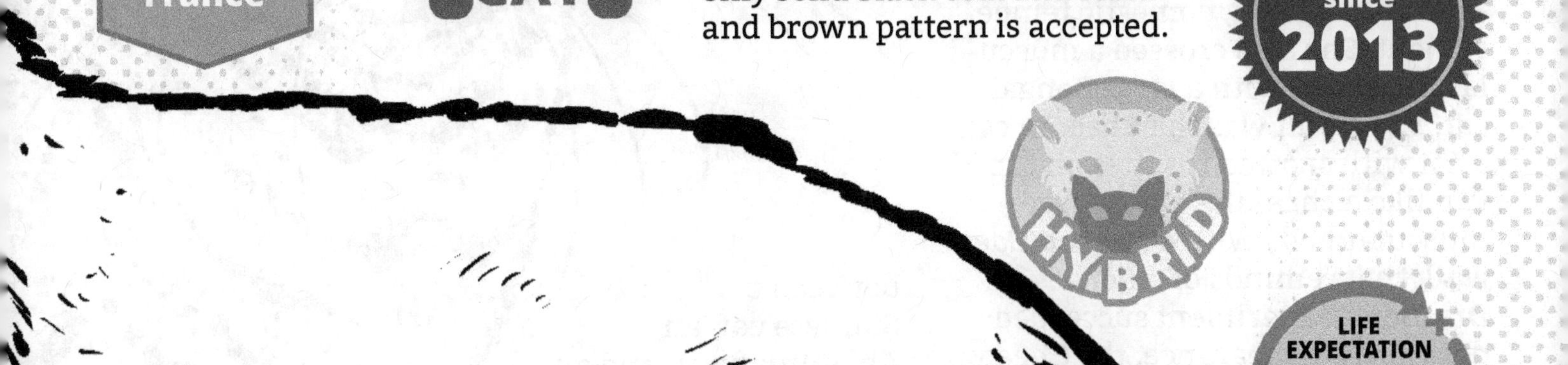

Market price:
400.00 USD–
3,500.00 USD

CHEETOH

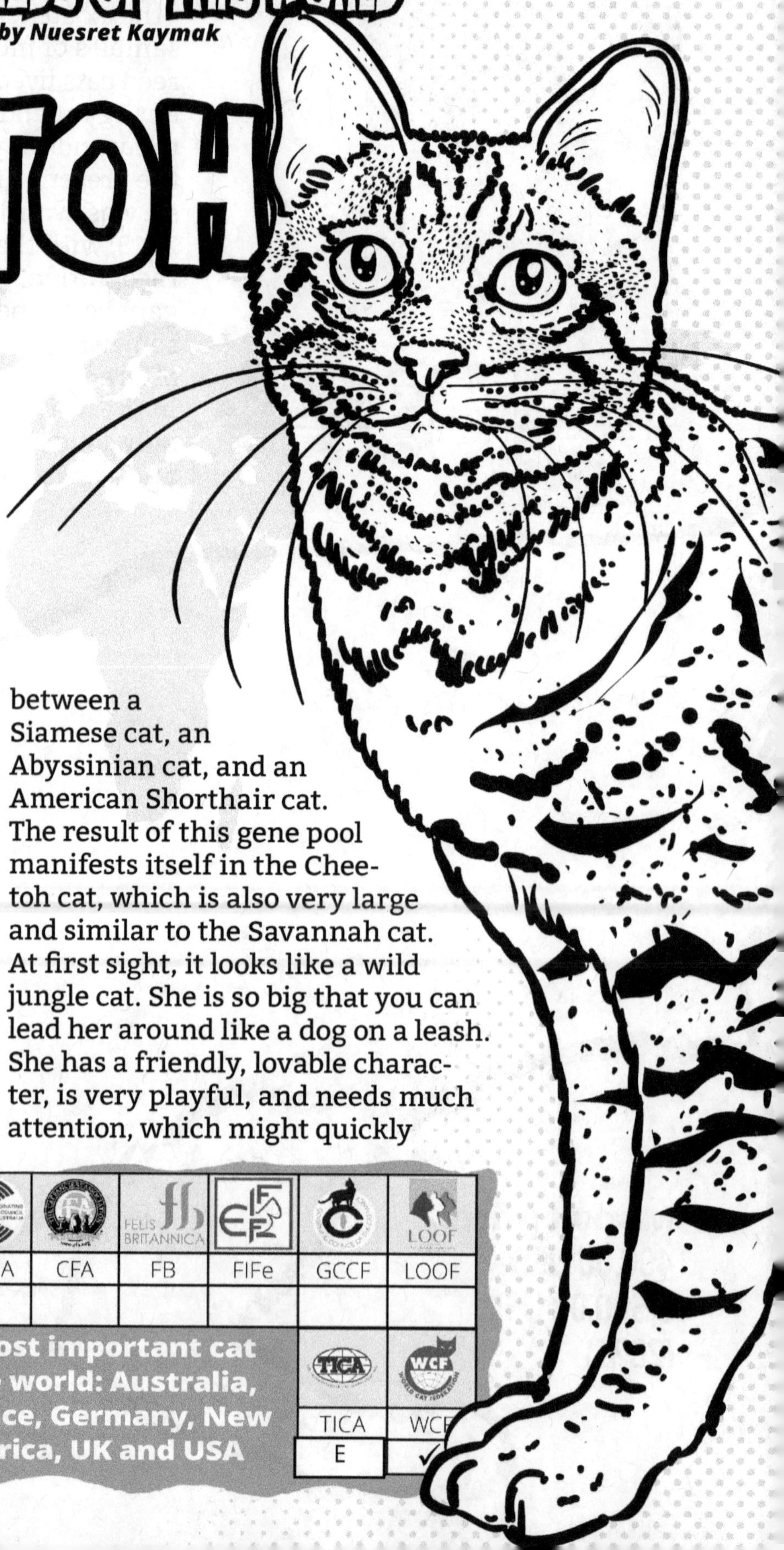

In 2001 the American cat breeder Carol Drymon started breeding a new cat breed. She wanted to breed a cat with a fearsome appearance - but with a calm, domestic temperament. She also crossed a muscular Ocicat cat with a large Bengal cat. The result was the Cheetoh cat: a cat with spotted rosettes on the fur, like a cheetah - and a large, muscular body with a low shoulder walk that reminds of a lion. Carol Drymon's experiment succeeded: despite his appearance, the Cheetoh was a peace-loving cat with a lovable nature. Of course, in the world of cat lovers, this was a bombshell. She quickly became world-famous and very popular. In 2004 TICA has registered the breed as an experimental new breed. As already mentioned, the Bengal cat comes from a cross between a wild Asian leopard cat and a domestic cat. The Ocicat cat, for her part, comes from a cross between a Siamese cat, an Abyssinian cat, and an American Shorthair cat. The result of this gene pool manifests itself in the Cheetoh cat, which is also very large and similar to the Savannah cat. At first sight, it looks like a wild jungle cat. She is so big that you can lead her around like a dog on a leash. She has a friendly, lovable character, is very playful, and needs much attention, which might quickly

ACF	ACFA	CAA	CCA-AFC	CCC	CCCA	CFA	FB	FIFe	GCCF	LOOF

NZCF	SACC	TICA	WCF
		E	✓

Selection of the most important cat associations in the world: Australia, Canada, China, France, Germany, New Zealand, South Africa, UK and USA

ATELIER KAYMAK
Science
PAPERS

BIG CAT

Recognized since 2004

Market price:
800.00 USD–
1,500.00 USD
(2023)

LUXUS

LIFE EXPECTATION 12-14 YEARS

overwhelm some owners. The playful and active cats, therefore, need much space for their run. An enclosure is advisable. Cheetoh cat is very sociable and due to its good-natured nature it gets along very well with children, dogs and other cats. The Cheetoh cat is bred today in the USA, Canada as well as in

Breeding start

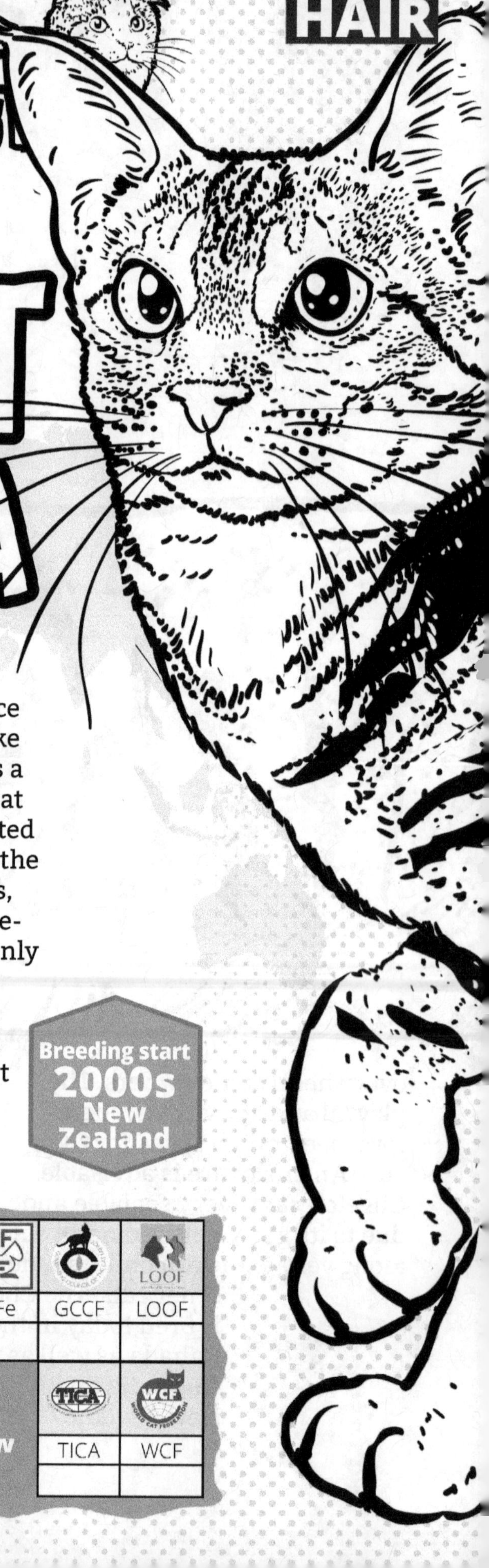

CLASSICAT ALSO JUNGALA

The athletically built Classicat, which is also called Jungala cat - is an offshoot of the Ocicat, but with classic tabby coloration. The cat is mainly bred in New Zealand, popular there under the name Jungala and has been bred there since the end of the 20th century. In neighboring Australia, they are known as Ocicat Classic cats. The name Jungala comes from Sanskrit and literally means wilderness - an allusion to the appearance of the cat, which looks like the wild Leopard cat. It is a derivative of the Ocicat cat - a cat breed that originated in 1964 in the USA from the gene pool of Siamese cats, Abyssinian cats, and American shorthaired cats. Only many decades later, one introduced the Ocicat in New Zealand - with what the native breeders didn't like the fur-pattern.

Breeding start
2000s
New Zealand

ACF	ACFA	CAA	CCA-AFC	CCC	CCCA	CFA	FB	FIFe	GCCF	LOOF

NZCF	SACC
✓	

Selection of the most important cat associations in the world: Australia, Canada, China, France, Germany, New Zealand, South Africa, UK and USA

TICA	WCF

So, they started to breed the Ocicat very cheekily. They crossed the fur pattern typical for wild cats and finally managed to give it its present appearance after a few litters. Since then, the Classicat, as it is now called, has enjoyed a growing number of followers throughout the Pacific. The New Zealand Cat Fancy Association recognized her in 2011.

Market price:
800.00 USD–
1,500.00 USD
(2025)

LIFE EXPECTATION 9-16 YEARS

Recognized since 2011

BIG CAT

Weight 16-32 lbs

CLIPPER CAT SHORTHAIR

Clipper cats were real superstars in the heyday of the English colonial era - which would have been unimaginable without the royal sea fleet. The cats suffering from polydactyly* because of a genetic defect could literally walk around doubly sure-footed on the rocking clipper boats, the English sailing ships, with the help of their handicap. The wooden ships were made for them because they could move on them - similar to professional mountaineers in their element - along with them. No wonder, then, that these cats were part of the regular crew as mice hunters on every more extended ship passage. They also enjoyed a reputation as lucky charms and were much appreciated by the captains as mascots. So, the cats with the broad paws finally reached Down Under - to the end of the world, as the South Pacific colonies were popularly called. An exceptionally high number of these cats succeeded in Australia and New Zealand in the context of the English colonization in the years from 1850 to 1900. So, one finds the highest density of polydactyly cats today on the New Zealand shore in Auckland and on the islands at the Hauraki Gulf. In countries like Germany, the Animal Welfare Act prohibits the breeding of Clipper Cats.

Breeding start
1950s
New Zealand

LIFE EXPECTATION
14-16
YEARS

ACF	ACFA	CAA	CCA-AFC	CCC	CCCA	CFA	FB	FIFe	GCCF	LOOF
							✓	✓		

NZCF	SACC	TICA	WCF

Selection of the most important cat associations in the world: Australia, Canada, China, France, Germany, New Zealand, South Africa, UK and USA

Weight
7-14
lbs

ORIGIN:

ENGLAND, NEW ZEALAND

Recognized since 1955

***Polydaktylie** is a condition in which a person is born with extra fingers or toes. The term comes from the Greek words for "many" ("*poly*") and "digits" ("*dactylos*").

Market price:
550.00 USD–1,200.00 USD (2025)

CLIPPER CAT LONGHAIR

In fact, the Clipper cat is a New Zealand cat. It is available as a short and as a Longhair cat. Their ancestors originally came from Europe - for example, from the warm southwest of the British Isles, which are spoiled by the warm Gulf Stream. Also, in Scandinavian countries, like Norway, one finds bone remains, which point to the existence of these mutated cats. In Europe, they were almost exterminated in the course of the Inquisition and the accompanying witch-hunt. Clipper cat cats are so-called polydactyly cats: a naturally occurring gene mutation causes polydactyly, multi-fingeredness, because of which the cats grow up to seven toes on each paw instead of five. At first sight, this may take some getting used to - but sometimes it gives the cat an enormous surefootedness - as, for example, in the snow (as with the lynx) and, of course, on wooden ships, their namesakes. In Germany, the Animal Welfare Act prohibits the breeding of Clipper cat cats.

Breeding start 1950s New Zealand

LIFE EXPECTATION 14-16 YEARS

Recognized since 1955

ACF	ACFA	CAA	CCA-AFC	CCC	CCCA	CFA	FB	FIFe	GCCF	LOOF
						✓	✓	✓		

NZCF	SACC	TICA	WCF

Selection of the most important cat associations in the world: Australia, Canada, China, France, Germany, New Zealand, South Africa, UK and USA

Weight 7-14 lbs

ORIGIN:

ENGLAND, NEW ZEALAND

Market price:
550.00 USD–
1,100.00 USD
(2025)

COLORPOINT SHORTHAIR

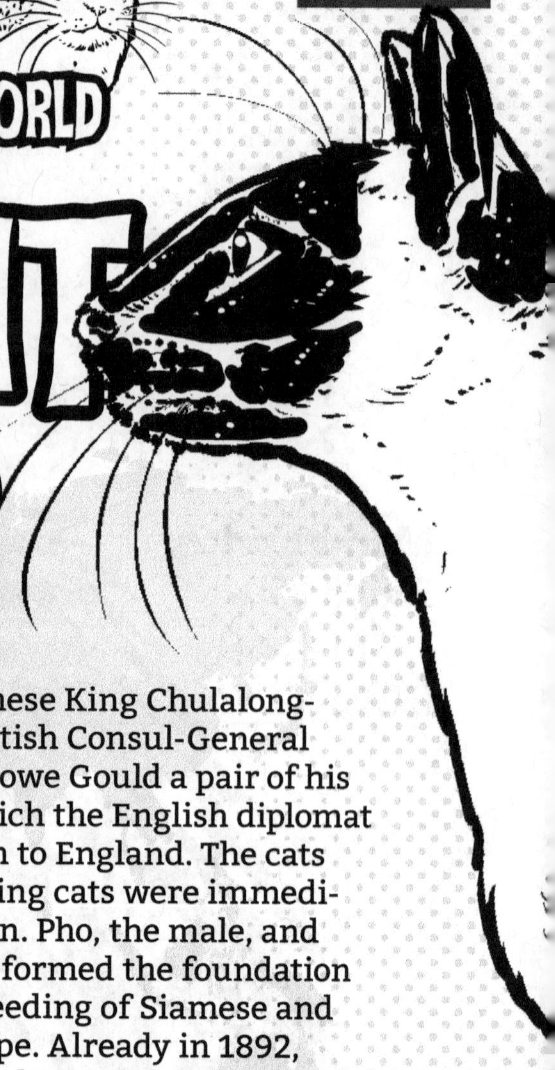

The Colorpoint Shorthair, aka Siamese, as they are called in some countries, is one of the most famous cats in the world. As the name suggests, the elegant cat comes from Southeast Asia - more precisely: from the Kingdom of Siam, today's Thailand. Scientists suspect that Arabian and Indian sailors brought the slim cats from the Mediterranean to Asia - where they mated with native cats and merged into them. Siamese cats are part-Albinos and have a unique pure-white short fur accordingly - that is colored at the running out extremities, the so-called Points - ears, tail, and paws - darkly until black. In the second half of the 19th century, travelers brought Siamese cats to Europe – where, unfortunately, most cats did not survive the travel strains. Therefore, they disappeared from the field of vision. In 1884, the Siamese King Chulalongkorn gave the British Consul-General Sir Edward Blencowe Gould a pair of his Siamese cats, which the English diplomat brought with him to England. The cats called Siamese King cats were immediately an attraction. Pho, the male, and Mia, the female - formed the foundation for the future breeding of Siamese and Thai cats in Europe. Already in 1892, they were certified as purebred cats and became a sensation. The Colorpoint Shorthair is like the Oriental Shorthair cat, a cross between a Siamese cat and an American Shorthair cat. During the 1940s, English and American cat breeders began to cross Siamese cats with other cats, which was an absolute taboo until then. But now, Siamese cats were crossed with Abyssinian cats and ordi-

ACF	ACFA	CAA	CCA-AFC	CCC	CCCA	CFA	FB	FIFe	GCCF	LOOF
			✓			✓				

NZCF	SACC	TICA	WCF

Selection of the most important cat associations in the world: Australia, Canada, China, France, Germany, New Zealand, South Africa, UK and USA

Breeding start
1940s
UK, United States

nary street cats. The declared aim was to breed other colors into the Siamese cat. Due to this outcrossing, it is not recognized by most federations, therefore as its own cat race. At best, it is classified into the color-class of the existing class of the Siamese cat in some associations. The organizations have - in order to maintain the standard - strict breeding conditions for Siamese cats. The Colorpoint Shorthair cat has the same color pattern as a Siamese cat but in entirely different colors: cream, lynx, red (tabby), and tortie. There are also many dot variations: Red dot, Cream Point, Cinnamon Point, Fawn Point, Seal Point, Chocolate Point, Blue Point, Lilac Point, Lynx Point, Tortie Point, and Torbie Point.

Market price:
100.00 USD–800.00 USD
(2023)

LIFE EXPECTATION 12-16 YEARS

Recognized since 1964

ANCIENT ORIGIN

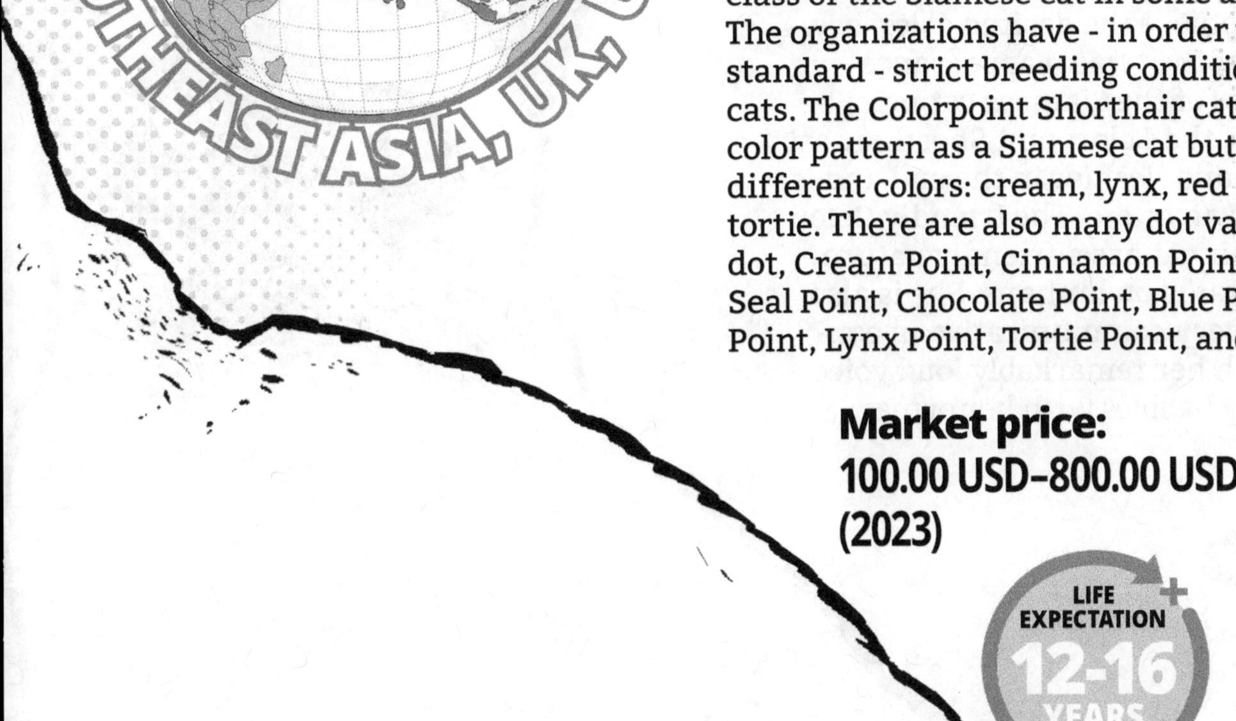

In order to achieve the desired color, however, some of the typical appearance characteristics of the Siamese cat must be omitted. Thus, the Colorpoint Shorthair cat looks heavier and fleshier in the end and is not so elegant anymore. The friendly Colorpoint Shorthair cat has a friendly nature and is very fixated on humans. She is also brilliant and very communicative in consequence: with her remarkably loud voice, she constantly babbles with her owner.

Traditional Siamese cat (Old-Style) with typical apple head

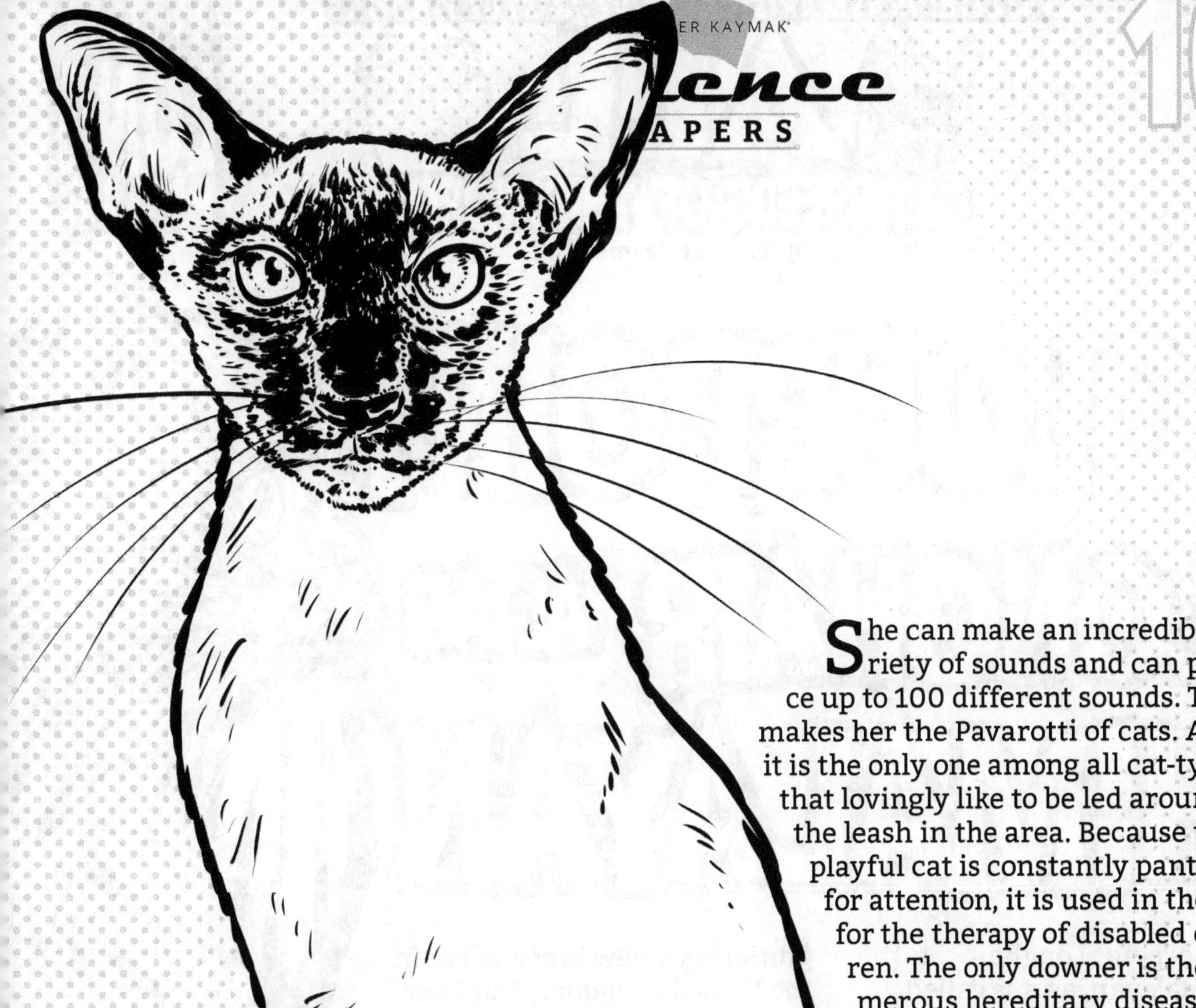

She can make an incredible variety of sounds and can produce up to 100 different sounds. This makes her the Pavarotti of cats. Also, it is the only one among all cat-types that lovingly like to be led around at the leash in the area. Because the playful cat is constantly panting for attention, it is used in the care for the therapy of disabled children. The only downer is the numerous hereditary diseases that make life difficult for the cat.

Modern Siamese cat with typical wedge head

COLOURPOINT LONGHAIR ALIAS HIMALAYAN

The Colorpoint Longhair cat, or Himalayan, as it is called worldwide - is a Siamese cat that has been crossed with a Persian cat. The latter she owes the beautiful coat, which ends in a bushy, shapely tail. She owes her reference to the Himalayas in her name to her beautiful, fluffy fur. Her breeding began in the 1920s in Europe and at the same time in the United States. In 1955 she was awarded recognition as a new breed of cat in the United Kingdom. The Colorpoint Longhair cat is a very social animal, and due to its friendly nature and play instinct, it is very suitable as a family cat.

Market price:
1,800.00 USD–3,500.00 USD (2025)

ACF	ACFA	CAA	CCA-AFC	CCC	CCCA	CFA	FB	FIFe	GCCF	LOOF
	✓	✓	✓							

NZCF	SACC	TICA	WCF
		✓	

Selection of the most important cat associations in the world: Australia, Canada, China, France, Germany, New Zealand, South Africa, UK and USA

ATELIER KAYMAK
ORIGIN:
ASIA, UK
Recognized since 1964
Breeding start 1940s UK, United States
LIFE EXPECTATION 12-16 YEARS
Autosomal domi-
nant polycystic

CORNISH REX

In 1950 an unusual kitten was found on a farm in the county of Cornwall, in the southwest of England. In the litter of Mrs. Nina Ennismore's farmer cat, there was a kitten with striking and very strange hair. The red tabby tom-cat wore a curled coat that looked like a blanket. Mrs. Ennismore was thrilled. It was a hair mutation due to the LPAR6 gene - the Lysophosphatidic acid receptor 6 - also called Rex Gene, which also occurs occasionally in humans. It is a genetic defect that triggers the wool hair syndrome and causes the premature termination of protein building in the hair. Mrs. Ennismore had bred animals with the same hair mutation in her rabbit breeding. The mole-like fur of the animals was needed for the production of fur coats. Coats made of sewn-together moleskins have always been trendy in the history of humanity. The coveted, expensive pelts have been in fashion for thousands of years. However, it was a costly affair to get the many moles together that were suitable for this. And for such a coat, one needed incredibly many pelts. The fur of Rex Rabbits came closest to the mole fur. Because the trade with the luxury industry was a lucrative business, there were at some point several Rex Rabbit breeders, as well as Mrs. Ennismore herself. She saw in the Rex kitten the potential to establish a new cat breed and started immediately.

ACF	ACFA	CAA	CCA-AFC	CCC	CCCA	CFA	FB	FIFe	GCCF	LOOF
✓	✓	✓	✓	✓	✓	✓	✓	✓	✓	✓

NZCF	SACC	TICA	WCF
✓	✓	✓	✓

Selection of the most important cat associations in the world: Australia, Canada, China, France, Germany, New Zealand, South Africa, UK and USA

She crossed the then sexually mature male with his mother and thus formed the foundation for the breeding of the so-called English-Rex cat. But the name was changed to Cornish Rex cat when soon after another and new Rex Variety appeared in the neighboring county of Devon - the Devon Rex cat. In 1956 the cat population of the farm had grown to 40 Cornish Rex cats. The problem was that there was no buyer. The lack of interest and the resulting lack of money caused Mrs. Ennismore to dispose of half of her cats. One of the surviving cats - Lamorna Cave - was sold to a breeder in the United States and shipped there. Lamorna Cave's Offspring were crossed with Siamese and other Asian cats, giving the Cornish Rex an elegant build. Cornish Rex cats are atypically small and very slim for English cats. Therefore, they don't do well in the typically cold and wet English winters. The breed was recognized by the American CFA in 1964 and by the British GCCF in 1967.

Market price:
200.00 USD–
2,500.00 USD
(2025)

Recognized since 1964

Breeding start
1950s
United Kingdom

Hypokaliämie, Progressive retinal atrophy (PRA), Baldness, Heart

CYMRIC

The Cymric cat is at home on the Isle of Man. It is the Longhair version of the Manx cat and, apart from the hair length, completely identical to it. Her name comes from the Celtic name for Wales, Cymru - which has absolutely nothing to do with her origin. She actually only got the name because it sounds very ancient, and you can distinguish it better from the Manx cat. The Cymric cat, like the Manx cat, has a short or no tail due to a gene mutation. Depending on the degree of mutation, there are four types of tail: rumpy (tailless), rumpy-raiser (tail with up to three bone vertebrae), stumpies (short, curly tail with kinks), and longies (normal looking tail). Longhaired kittens were usually undesirable in litters of Manx cats – until a shrewd Canadian breeder, Althea Frahm, made them known as Manx mutants in the 1960s. The Cymric cat is recognized by some international cat associations. However, some other associations see it only as a variant of the Manx cat. One cat association amusingly describes its appearance as a living bowling ball that whizzes through the neighborhood. The intelligent Cymric has no fear of the water and is very playful. She is very agile and, therefore, a real acrobat. She gets along very well with people and other pets and is a good family animal. The gene mutation causes the so-called Manx syndrome, which causes incurable disabilities in many Cymric cats. In countries such as Germany, the Animal Welfare Act prohibits the breeding of Cymric cats.

BOBTAIL CAT

LIFE EXPECTATION 8-14 YEARS

Market price:
500.00 USD–800.00 USD (2023)

ACF	ACFA	CAA	CCA-AFC	CCC	CCCA	CFA	FB	FIFe	GCCF	LOOF
							✓	✓		

NZCF	SACC	TICA	WCF
			✓

Selection of the most important cat associations in the world: Australia, Canada, China, France, Germany, New Zealand, South Africa, UK and USA

ATELIER KAYMAK
Science
PAPERS
ORIGIN:
ISLE OF MAN, EGYPT, MIDDLE EAST
Recognized since 1994
Breeding start 1960s Canada

SHORT HAIR

DEVON REX

Breeding start 1868 United Kingdom

REX CAT

The Devon Rex cat was the second Rex cat in England to be discovered just a few years after the Cornish Rex cat. The peculiar cat, often referred to as the Goblin or Alien cat because of its wedge-shaped head and pronounced nose, was found in 1959 by Beryl Cox in Buckfastleigh - a former small water mill town in the county of Devonshire. The puppy, which was initially thought to be a Cornish Rex cat, had a genetically completely different mutation - instead of the LPAR6 gene of the Cornish Rex cat - the gene KRT71 and encodes the keratin-71 factor. This gene defect, which occurs earlier in Rex rabbits, manifests itself in the form of particularly short, straight hairs, which grow together to form a dense, shiny coat and make it look like a mole fur. The Devon Rex cat is a faithful soul and follows its owner at every turn. She wants constant body contact, although she is by no means sluggish. On the contrary: she is so agile and playful that she is called a monkey disguised as a cat in English-speaking countries! She jumps around a lot and climbs all the viewpoints in the apartment. Therefore, she loves to sit on the shoulders of her keeper.

ACF	ACFA	CAA	CCA-AFC	CCC	CCCA	CFA	FB	FIFe	GCCF	LOOF
✓	✓	✓	✓	✓	✓	✓	✓	✓	✓	✓

NZCF	SACC	TICA	WC
✓	✓	✓	✓

Selection of the most important cat associations in the world: Australia, Canada, China, France, Germany, New Zealand, South Africa, UK and USA

Her curiosity is so great that she explores every cavity she finds and sounds it out with her body. You can teach her tricks and walk her on a leash. No wonder that comparisons with dogs and even parrots do not fail. The only downer: the Devon Rex cat lacks the vibrissae, the tactile hairs. Thus the cat lacks the species-typical and decisive help for orientation and communication. In countries such as Germany, the Animal Welfare Act prohibits the breeding of the Devon Rex cat.

Market price:
800.00 USD–
4,000.00 USD
(2025)

Recognized since **1882**

LIFE EXPECTATION **14-16** YEARS

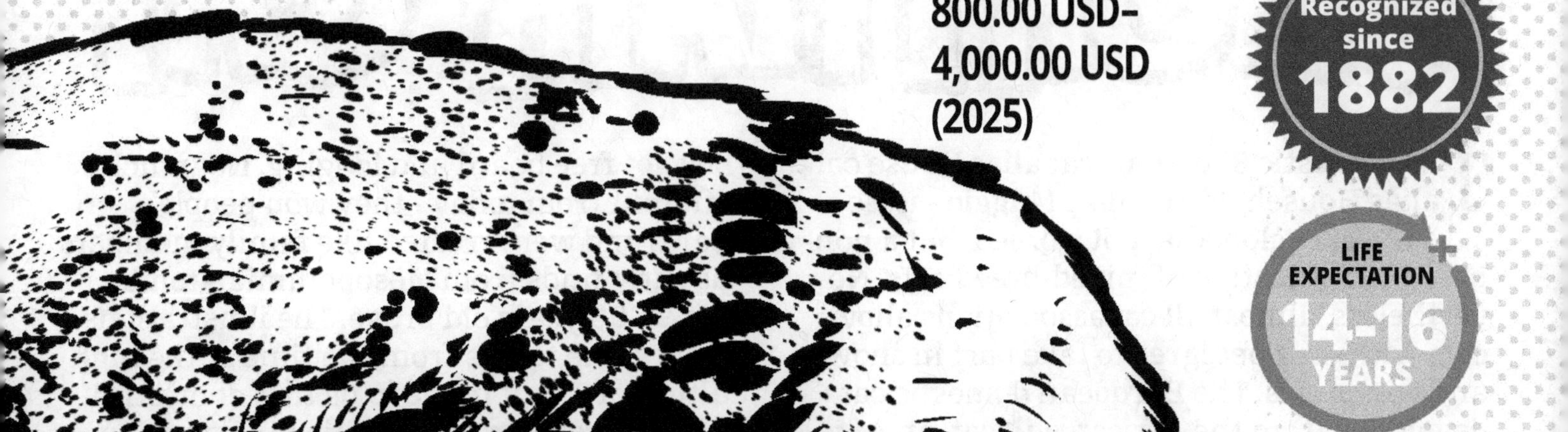

(with pure white coat): congenital hearing loss, deafness, disturbances of the sense of balance, Eye tremor, muscle diseases, potassium deficiency

SHORT HAIR

ANCIENT ORIGIN

DOMESTIC SHORTHAIR CAT

The domestic Shorthair cat alias House cat alias Household Pet alias Muggie - is, so says the encyclopedia: not a breed, but a non-breed classification of mixed-breed cats. Nevertheless, almost all cat associations allow cats without a pedigree to take part in shows and win prizes. The European domestic cat developed from the African wildcat, an African wild cat that first appeared in Africa 130,000 years ago. When humans changed from nomads to farmers, they began to settle down. They established cultural spaces that quickly grew into larger communities, settlements, and villages. Cats were there right at the beginning. They were popular and therefore liked to be close to people with whom they lived in the symbiotic community: for protecting their food from mice and rats - they got free board and lodging. Not much different from today - they won people's hearts and were treated like family members - as early finds from Mesopotamian tombs over 9,500 years old prove. The illustrations on old stone tombs from the time of the high culture of Egypt prove to us that domesticated cats lived and were deified in harmony with humans. A regular cat cult existed, and even a cat goddess - Bastet, daughter of the sun god Ra - who paid homage to the Egyptians. Everything revolved around cats; they were worshipped and protected. Cat torturers and cat murderers were themselves punished with death. The Egyptians even lost battles against the attacking Persians, who were not too good at using sacred

known since
9500 v. Chr.
Asia

ACF	ACFA	CAA	CCA-AFC	CCC	CCCA	CFA	FB	FIFe	GCCF	LOOF
								✓		

NZCF	SACC	TICA	WCF
		✓	

Selection of the most important cat associations in the world: Australia, Canada, China, France, Germany, New Zealand, South Africa, UK and USA

Weight
7-9
lbs

NON-BREED CLASSIFICATION

LIFE EXPECTATION
12-18
YEARS

Gangliosidosis

cats as projectiles against the Egyptian army to demoralize and paralyze them. The trade with cats was strictly controlled by priests, and exporting them to foreign countries was practically impossible. Still, Phoenician traders and Romans managed to smuggle cats out of Egypt to Europe on ships secretly. There, these cats formed then the basis of today's domestic cats.

Market price:

DON SPHYNX ALIAS DONSKOY

The wrinkled bare-cat is born either already completely bare or loses its fur at the latest in the sexually mature one. The cat breed has a tragic origin, which could come directly from the imagination of a Russian playwright of the 19th century. In 1987, on her way home in Rostov-on-Don, a southern Russian metropolis on the Assow Sea, Professor Elena Kovaleva made a creepy discovery: in a playground, children threw at each other a plastic bag from which the miserable sounds of a cat were coming. Swearing, she abruptly ended the animal cruelty and took the bag from the children, in which there was an unfortunate little tortie-coloured kitten. Apparently someone had wanted to get rid of the cat, put it into the bag and threw it into the river. The children must have found her immediately afterwards and pulled her out of the water. Mrs. Kovaleva didn't think long and took the kitten home. The professor already owned some cats, but had no problem to take up another one. After some time Varvara, as the cat was now called, lost all her hair - which caused the scared Elena to rush to the next vet. The doctor could not diagnose any illness and was at a loss. As it turned out, Varvara's hair loss was a genetic defect, a random naked cat mutation. The Don Sphynx cat was recognized as a new breed by the WCF in 1997. In countries like Germany the Animal Welfare Act prohibits the breeding of the Donskoy cat because of the missing vibris, the tactile hairs.

HAIRLESS CAT

ACF	ACFA	CAA	CCA-AFC	CCC	CCCA	CFA	FB	FIFe	GCCF	LOOF
			✓	✓			✓	✓		✓

NZCF	SACC	TICA	WCF
		✓	

Selection of the most important cat associations in the world: Australia, Canada, China, France, Germany, New Zealand, South Africa, UK and USA

LUXUS

ORIGIN:
ASIA, SOVIET UNION
Market price:
1,800.00 USD–
5,000.00 USD
(2025)
Recognized since 1997
Breeding start
1987
Soviet Union
LIFE EXPECTATION
12-15
YEARS
muscular disorders

SHORT HAIR

Breeding start
2004
China

DRAGON LI ALIAS CHINA LI HUA CAT

The Dragon Li cat, or China Li Hua cat, probably comes from the Chinese wild cat (Felis silvestris bieti). The thesis is still controversial here and there - but it could not be refuted scientifically so far. This makes the Dragon Li cat something exceptional because almost all other cats originate from the African wild cat (Felis silvestris lybica) or the European wild cat (Felis silvestris silvestris). The Dragon Li cat has a tabby pattern and a fleshy and muscular build. The first Dragon Li cat - a tomcat named Needy - was presented to

ACF	ACFA	CAA	CCA-AFC	CCC	CCCA	CFA	FB	FIFe	GCCF	LOOF
		✓								

NZCF	SACC	TICA	WCF

Selection of the most important cat associations in the world: Australia, Canada, China, France, Germany, New Zealand, South Africa, UK and USA

the public in 2004 in China's capital Beijing. Only one year later, in 2005, the Dragon Li cat was recognized as a new cat breed by the China cat Aficionado Association CAA, the Chinese branch of the American cat Fanciers' Association CFA.

LIFE EXPECTATION
9-16
YEARS

Recognized since
2005

Weight
4,5-6
lbs

Market price:
400.00 USD–
1,000.00 USD
(2023)

EGYPTIAN MAU

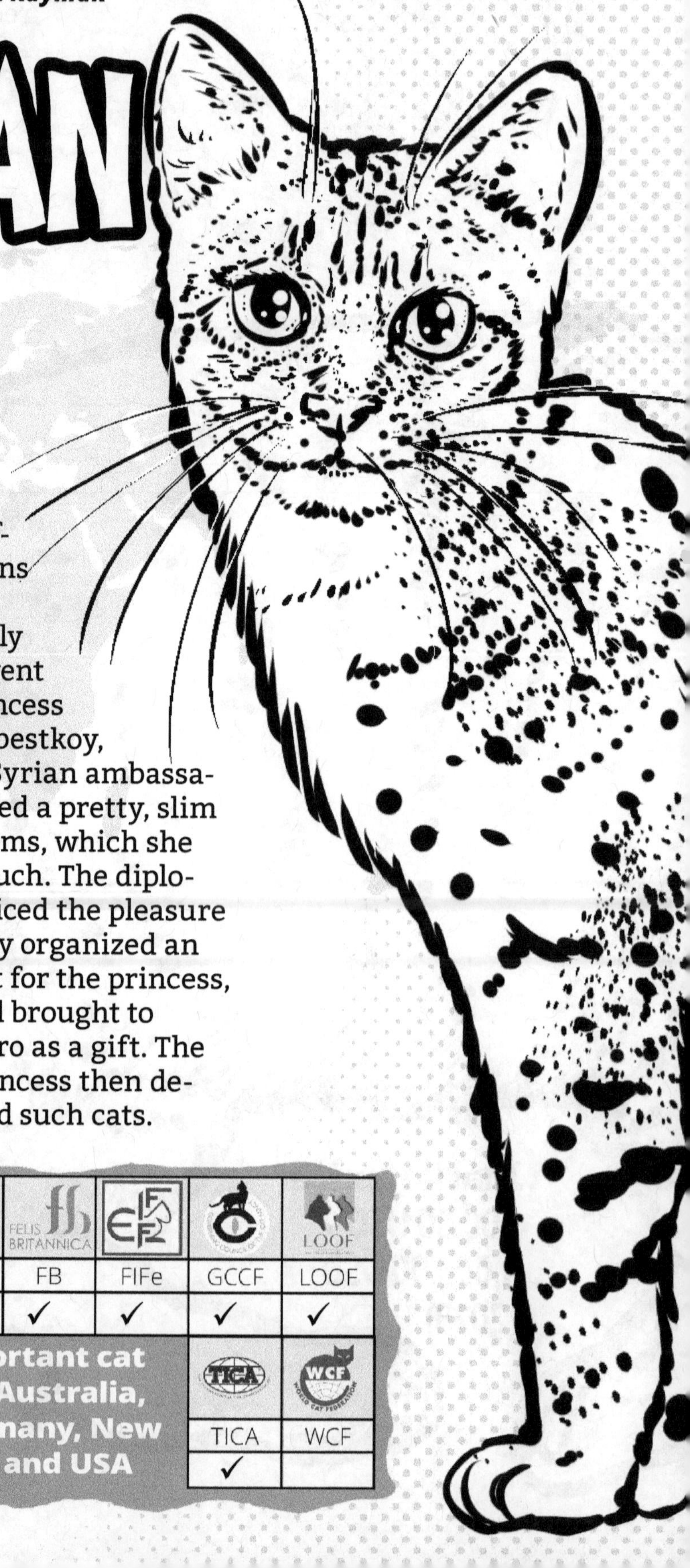

No civilization loved the cat more than the high culture of Kemet - the Black Land - as ancient Egypt is called. The name refers to the black mud which, after the annual Nile -flood, brought the farmers a productive harvest and well-filled grain stores. Cats played an enormously important role at this time - because they protected the valuable grain stores of the Egyptians from pests such as mice and rats. The exciting story about the origin of the modern Egyptian Mau cat can be traced back to a Russian princess who lived in exile in Rome after the Second World War.

There are different versions of this story, but essentially it probably went like this: Princess Natalie Troubestkoy, visiting the Syrian ambassador, discovered a pretty, slim cat in his rooms, which she liked very much. The diplomat who noticed the pleasure - immediately organized an Arab Mau cat for the princess, which he had brought to her from Cairo as a gift. The delighted princess then decided to breed such cats.

ACF	ACFA	CAA	CCA-AFC	CCC	CCCA	CFA	FB	FIFe	GCCF	LOOF
✓	✓	✓	✓	✓	✓	✓	✓	✓	✓	✓

NZCF	SACC	TICA	WCF
✓	✓	✓	

Selection of the most important cat associations in the world: Australia, Canada, China, France, Germany, New Zealand, South Africa, UK and USA

The female, called Baba by the princess - together with the tomcat of the diplomat, Geppo, formed the basis for the breeding program of the modern Egyptian Mau cat. In 1956 the princess moved with her cats to the United States and two years later, in 1958, registered her cats as a new cat breed Fatima. In 1968 the Egyptian Mau finally received the championship status of the CFF. The recognition of the CFA as a separate breed did not take place until 1977. The attempt to breed the Egyptian Mau in Great Britain was unsuccessful - but formed the basis for the breeding of the Ocicat. Since 1988, the Egyptian Mau has also been bred in Switzerland with the help of animals imported from the USA.

Market price:
200.00 USD–3,000.00 USD (2025)

Recognized since 1968

LIFE EXPECTATION 18-20 YEARS

Breeding start 1950 United States

ATELIER KAYMAK'S CREATIVE- & STUDY BOOKS

ALL CAT BREEDS OF THIS WORLD

by Nuesret Kaymak

SHORT HAIR

Breeding start 2004 United States

ELF CAT

In the middle of the 2000s, a new cat breed was born in the USA: The Elf cat. Karen Nelson and Kristen Leedom had previously crossed an American Curl cat, a cat with backward-curved roll ears - with a Canadian Sphynx, a bald cat with a recessive gene defect. After unsuccessful years and some frustration - they finally succeeded in the experiment. The playful elf-cat has a muscular body with many skin folds. The athletically built climbing artist weighs unbelievably, often up to 7 kg. The eye color of the heavyweight varies between blue, green, or both (odd-eyed) - can also be yellow. The friendly elf-cat is very affectionate and in need of love. Therefore, she enjoys the closeness of humans but also that of other cats. The hairlessness has its pitfalls: thus, the Elf-cat is exposed to the weather without protection and can suffer from skin diseases fast with insufficient care. In countries like Germany, the Animal Welfare Act prohibits the breeding of the elf-cat.

HAIRLESS CAT

LIFE EXPECTATION 12-14 YEARS

Market price:
2,000.00 USD–
2,500.00 USD
(2023)

ACF	ACFA	CAA	CCA-AFC	CCC	CCCA	CFA	FB	FIFe	GCCF	LOOF

NZCF	SACC	TICA	WCF
		✓	

Selection of the most important cat associations in the world: Australia, Canada, China, France, Germany, New Zealand, South Africa, UK and USA

Weight 10-15 lbs

ATELIER KAYMAK
Science
PAPERS
ORIGIN:
UNITED STATES
Recognized since 1968

Breeding start
1870s
United Kingdom

EUROPEAN BURMESE

Diabetes mellitus, Hypokalaemia, Pica, Hypokalaemia, Gangliosidosis, Flat-chested Kitten Syndrom (FCKS),Problems with eyes, kidneys (stones) and sensory organs (hypaesthesia)

The old Burmese cat, known as Su-Pa-Lak or Thong Daeng, has been worshipped as a holy cat since the 14th century in the Buddhist temples of the Thai-Burmese border areas, where priests were assigned to care in training. At that time, cats were a central part of Southeast Asian culture. The old Thai book of the cat poems Tamra Maew from this time reads: Of splendid appearance and the best form - dyed like copper, this cat is beautiful - the light of her eyes is like a shining ray. In today's Thai cat shows, they distinguish between the native Suphalak and the Burma cat, which were later developed in the West. Long before the modern Burmese cat was developed as a breed in the United States, there were records of dark copper Siamese cats in cat shows in 19th century Britain. However, the darker color variations fell out of favor with associations and breeders compared to their lighter cousins. This only changed in 1952, when the breeders Sydney and Lillian France imported three American Burmese cats to Great Britain, and these then formed the basis for a new European breed of Burmese cats. With success: the English GCCF recognized it in the same year. It is known that the European Burma cat has more of the original Asian characteristics than the American Burma cat. This is expressed by the almond-shaped eyes, longer, wider set ears, and the Variety of coat colors.

ACF	ACFA	CAA	CCA-AFC	CCC	CCCA	CFA	FB	FIFe	GCCF	LOOF
✓	✓	✓	✓	✓	✓	✓	✓	✓	✓	✓

NZCF	SACC	TICA	WCF
✓	✓	✓	✓

Selection of the most important cat associations in the world: Australia, Canada, China, France, Germany, New Zealand, South Africa, UK and USA

ATELIER KAYMAK
Science
PAPERS
Recognized since 1952
ORIGIN:
SOUTH ASIA, ENGLAND
LIFE EXPECTATION
16-18
YEARS
Market price:
600.00 USD–
2,000.00 USD
(2023)
Weight
6-10
lbs

Breeding start
1960s
United States

EXOTIC
SHORTHAIR

The exotic Shorthair cat has been around since the middle of the 20th century. Persian cats had always been very popular in the USA - however their attitude was very demanding. The cats demanded a lot of time from their owner for grooming - if he didn't want his favourite to turn into a matted ball on four legs. Many cat lovers therefore secretly wanted a cat that was easier to care for. They wanted a Persian cat with all the positive qualities - but with a coat they didn't have to do anything else for. So cat breeders experimented with Persian cats, Burmese and American cats as well as British Shorthair cats. They finally bred a Persian cat - one that lived up to this wish. The cat with the short, plush fur was called immediately in the vernacular the lazy man's Persian. But the cat ultimately achieved world fame in a totally different way: it was the model for one of the most famous comic strip characters in the world: a red tiger cat was the inspiration for comic artist Jim Davies - who has been portraying his tomcat Garfield since the end of the 1970s. A fat, lazy cat who prefers to eat the day over lasagna and expresses himself sarcastically to God and the world.

ACF	ACFA	CAA	CCA-AFC	CCC	CCCA	CFA	FB	FIFe	GCCF	LOOF
✓	✓	✓	✓	✓	✓	✓	✓	✓	✓	✓

NZCF	SACC	TICA	WCF
✓	✓	✓	✓

Selection of the most important cat associations in the world: Australia, Canada, China, France, Germany, New Zealand, South Africa, UK and USA

ATELIER KAYMAK

Science
PAPERS

ORIGIN:

ASIA, UNITED STATES

Market price:
200.00 USD–
5,500.00 USD
(2025)

Recognized since 1967

TOP 10

LIFE EXPECTATION 10-15 YEARS

Autosomal dominant Polycystic kidney disease (ADPKD); (with pure white coat) Congenital hearing loss, deafness, disturbances of the sense of equilibrium, Eye tremor

SHORT HAIR

Breeding start
1990s
UK, Canada

FOLDEX SHORTHAIR

In 1993, a particularly lovable cat was discovered in a cat show in Québec, Canada. The US-American Betty-Ann Yaxley presented a specimen of her new breed, which immediately attracted the attention of all visitors. The sweet cat had a peculiar and winning appearance because it came out from a cross of a Scottish fold cat with an Exotic Shorthair cat. For some, she looked like a cuddly Persian with short, dense fur and slightly deformed ears. Others thought it was the cheerful version of Jim Davies tomcat Garfield with folding ears. There were also comparisons with teddy bears. So much concentrated affection and admiration hardly any other cat in the world could enjoy. The breeder Jeanne Barrette fell in love with this cat so much that she was breeding the FoldEx cat full-time and developed the breed further.

ACF	ACFA	CAA	CCA-AFC	CCC	CCCA	CFA	FB	FIFe	GCCF	LOOF
			✓	✓						

NZCF	SACC	TICA	WCF

Selection of the most important cat associations in the world: Australia, Canada, China, France, Germany, New Zealand, South Africa, UK and USA

Thanks to the joint efforts of Yaxley and Barrette, the FoldEx cat received experimental CCA status three years after its debut and full championship status in 2010. Since only the Canadian CCA and also in Canada resident CCC recognizes the race cat - the FoldEx cat is to be found very rarely outside Canada. In countries like Germany, the Animal Welfare Act prohibits the breeding of the FoldEx cat.

Market price:
500.00 USD–
1,500.00 USD
(2023)

Recognized since 2010

LIFE EXPECTATION 12-15 YEARS

Susceptible to joint diseases

DESIGNERCAT

Breeding start
1990s
Canada

FOLDEX LONGHAIR

The FoldEx cat, which originates from Canada, is a cross between the Exotic Shorthair cat and the Scottish Fold - the Scottish fold-eared cat. The FoldEx cat resembles teddy bears due to its large round eyes, thick head and dense fur. The nose is shorter than that of a Scottish Fold-eared cat - but not quite as short as that of the Exotic Shorthair cat. In temperament, the intelligent FoldEx is very affectionate and sociable. She is, to say it casually, a delight for every cat lover. She has an unbelievable amount to tell and thus absolutely does not belong to the introverted or cranky cats. Gripped by curiosity, the playful one explores every spot of her home and romps around the apartment all day long.

DESIGNERCAT

ACF	ACFA	CAA	CCA-AFC	CCC	CCCA	CFA	FB	FIFe	GCCF	LOOF
			✓	✓						

NZCF	SACC	TICA	WCF

Selection of the most important cat associations in the world: Australia, Canada, China, France, Germany, New Zealand, South Africa, UK and USA

ORIGIN:

UNITED KINGDOM, CANADA

The FoldEx cat will climb anything that looks high and wobbly enough for daring experiments and adventurous expeditions. Malicious tongues claim that in her euphoric play she resembles a mop that flits back and forth through the area.

Market price:
500.00 USD–
1,500.00 USD
(2023)

LIFE EXPECTATION
12-15
YEARS

Recognized since
2010

Susceptible to joint diseases

Weight
6-14
lbs

FOREIGN WHITE SHORTHAIR

The foreign white cat is the pure white version of the Siamese cat. The history of this young cat breed began in 1962 when the English cat breeder Patricia Turner began to cross Siamese cats with light blue eyes and white domestic cats with green eyes. Their intention was to breed the bright blue eyes of Siamese cats to white domestic cats - but to spare the cats from the diseases usually associated with them. Because white cats with light blue eyes generally have hereditary diseases such as deafness - and suffer from vision disorders. But this is precisely what Mrs. Turner wanted to breed away. A few years later, she actually succeeded in this plan, and a healthy foreign white cat, as she was now called, saw the light of day to the delight of Mrs. Turner. In the year 1966, the foreign white cat was exhibited for the first time and immediately found international attention. Since then, the beautiful blue eye color of this elegant cat has been called Turner Blue. In 1977 the Foreign White cat was finally recognized as a new cat breed.

ACF	ACFA	CAA	CCA-AFC	CCC	CCCA	CFA	FB	FIFe	GCCF	LOOF
✓					✓					

NZCF	SACC	TICA	WCF

Selection of the most important cat associations in the world: Australia, Canada, China, France, Germany, New Zealand, South Africa, UK and USA

ORIGIN:
SOUTHEAST ASIA, UK
Breeding start
1950s
United States
Recognized since
1882
Market price:
600.00 USD–
3,000.00 USD
(2023)
(with pure white coat): congenital hearing loss, deafness, disturbances of the sense of balance, Eye tremor
LIFE EXPECTATION
15-20
YEARS

FOREIGN WHITE LONGHAIR

One of the most beautiful descriptions of the long-haired foreign white cat - also called Chantilly - reads: Her fur is like freshly fallen snow and her eyes shine like fresh, blue glacier meltwater. The foreign white longhaired cat was first introduced in the 1970s and arose from the desire of many cat lovers for a longhaired variant of this elegant and very coveted cat. Unfortunately, every cat that has a dominant white coat - i.e., an albino - has a genetic defect that is usually accompanied by deafness. The cause is an albinism enzyme, which only reacts at temperatures above 37 degrees - which corresponds to the body temperature of mammals - and thus prevents the roots of the follicles from producing melanin, a pigment. This means, for example, that in vitro kittens carrying the gene react to the warmth of their mothers' wombs, and the puppies are born with a bright white coat.

Breeding start 1969 United States

LIFE EXPECTATION 15-20 YEARS

Weight 8-12 lbs

ACF	ACFA	CAA	CCA-AFC	CCC	CCCA	CFA	FB	FIFe	GCCF	LOOF
✓					✓					

NZCF	SACC	TICA	WCF

Selection of the most important cat associations in the world: Australia, Canada, China, France, Germany, New Zealand, South Africa, UK and USA

Recognized since 1979

Market price:
800.00 USD–
1,500.00 USD
(2023)

(with pure white fur): Congenital

GERMAN LONGHAIR

Until the 1950s, white Longhair cats were commonly called Angora cats. The term Angora refers to a former province in the Ottoman Empire (today Turkey), where it is found in the name of the capital, Ankara. The province of Angora was famous for its white-haired cats, rabbits, and goats with fluffy fur. The German Longhair cat has a similar appearance to the British Longhair cat and shares the heritage with the Persian cat. Longhaired cats were bred in Germany according to colors rather than strict pedigree standards and were simply called Angoras. The German biologist Dr. Friedrich Schwangart founded the cat breed in a somewhat derogatory way. Because he criticized - entirely according to the prevailing zeitgeist - the meanwhile mixed breeding line of the Deutsch Longhair cat - in comparison to the relatively pure pedigree of the British Persian cat, he suspected that the elongated face of the Deutsch Longhair cat was the result of crossbred genes from Scandinavian wild cats. From 1929 to 1932, Dr. Schwangart meticulous-

Breeding start
1920s
Germany

ACF	ACFA	CAA	CCA-AFC	CCC	CCCA	CFA	FB	FIFe	GCCF	LOOF

NZCF	SACC	TICA	WCF
			✓

Selection of the most important cat associations in the world: Australia, Canada, China, France, Germany, New Zealand, South Africa, UK and USA

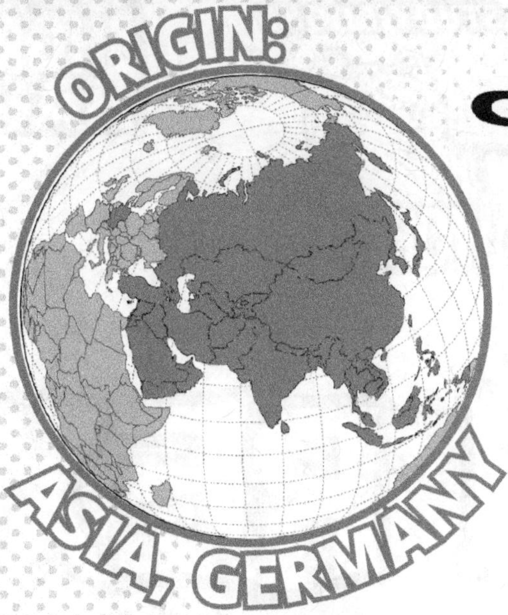

Market price:
100.00 USD–1,200.00 USD (2023)

ly recorded the pedigree of his German Longhair cats like a Dr. Frankenstein in a breeding booklet in which he also dealt with breeding techniques of his new breed. The Second World War interrupted Schwangart's breeding program and destroyed his records, with the result that the German Longhair cat was forgotten entirely. In 1968 some still living specimens of German Longhair cats were found in the estate of a certain Mrs. R. Aschemeier - which the lady had apparently rescued during the war - and brought into her care. An official revival of the cat breed finally took place in the 2000s. The Deutsch Longhair cat was fully recognized by the WCF in 2012.

Recognized since 2012

LIFE EXPECTATION

GERMAN REX

According to old photos, there were German Rex cats apparently already in East Prussia of the 1930s. Thus, a strange tomcat Munk allegedly lived with Mrs. Erna Schneider in the city of Königsberg, today's Kaliningrad. The tomcat Munk is said to have been the result of crossing an Angora cat with a Russian blue. As already explained in the section of the Cornish Rex cat, Rex cats have a unique gene defect, which causes crimped, wool-like hair growth. However, the official beginning of the German Rex breed did not begin until 1951, when Dr. Rose Scheuer-Karpin walked through the park of the Hufeland Hospital, a former nuthouse in Berlin-Buch in the district of Pankow. While she was strolling in the park to relax, Mrs. Scheuer-Karpin suddenly came across a tiny, lovely kitten that immediately approached her. Mrs. Scheuer- Karpin had instantly fallen in love with the animal and decided to take it in without further ado. She inquired at the institution about the fund and learned that the kitten was called Lamby (Lämmchen) and was the darling of all employees and patients. However, this obviously did not prevent Mrs. Scheuer-Karpin from taking the cat with her. The busy owner exhibited her cat everywhere and managed to have Lamby internationally recognized in the same year of her discovery as German Rex and thus as her own cat breed, right after the Cornish Rex cat. Six years later, in 1957, Lamby gave birth to Rex Offspring. This litter became the foundation for the breeding of the German Rex cat as it is known worldwide today.

Breeding start
1950s
Germany

ACF	ACFA	CAA	CCA-AFC	CCC	CCCA	CFA	FB	FIFe	GCCF	LOOF
							✓	✓		

NZCF	SACC	TICA	WCF
			✓

Selection of the most important cat associations in the world: Australia, Canada, China, France, Germany, New Zealand, South Africa, UK and USA

ATELIER KAYMAK
Science
PAPERS
ORIGIN:
MIDDLE EAST, GERMANY
Recognized since 1979
LIFE EXPECTATION
9-14
YEARS
Market price:
500.00 USD–
1,000.00 USD
(2023)
Weight
5-8

SHORT HAIR

DESIGNERCAT

HAVANA ALSO HAVANA BROWN

Actually, one would think that this beautiful chestnut-colored cat had emerged from the thick cigar smoke of Cuba. Instead, it was developed soberly in 1952 in England. The story happened as follows: in 1952, Mrs. Munro-Smith was quite surprised when she discovered many chestnut kittens in her cat's litter. This was way stranger because the offspring came from a cross between a Colorpoint cat and a Persian cat, and such a connection should not have produced brown kittens. What happened afterward had far-reaching consequences. Mrs. Munro-Smith kept some of the puppies and sent the rest of the kittens to an American breeder in California. This led to a divergence between the American Havana Brown cat and the British Havana cat. While the American breeding program produced chestnut-colored Havana Brown cats - the British version is more oriented towards slim Siamese cats with color range.

Breeding start
1950s
United States

ACF	ACFA	CAA	CCA-AFC	CCC	CCCA	CFA	FB	FIFe	GCCF	LO
	✓	✓	✓	✓		✓				✓

NZCF	SACC	TICA	WCF
		✓	

Selection of the most important cat associations in the world: Australia, Canada, China, France, Germany, New Zealand, South Africa, UK and USA

ATELIER KAYMAK
Science
PAPERS
ORIGIN:
ASIA, UNITED STATES
Recognized since 1964
Market price:
500.00 USD–
1,300.00 USD
(2023)
LIFE EXPECTATION
12-15
YEARS

HIGHLAND FOLD

The Highland Fold cat is the long-haired version of the Scottish folding ear cat - into which the genes of the long-haired Persian were crossed. After the controversy in Europe about the health aspects of breeding folding ear cats started and the breeding ban was imposed, it moved to the USA. There, US-American cat breeders began to continue, which was forbidden to the European breeders. This worked because the Highland Fold cat had not lost any of its popularity despite the quarrels. In 1980 the cat shows organized in the USA by Mrs. Hazel Swadberg made the Highland Fold cat popular. Six years later, TICA recognized it as an independent breed. Obviously, they could successfully eliminate the genetic factors for the typical, suffering hereditary diseases, since apart from the characteristic of the race-typical ears - otherwise healthy puppies came into the world. However, there are still very few breeders, which has a strong effect on supply and demand. In Germany, the breeding of folding ear cats is not allowed because of the animal protection law.

Breeding start
1950s
United Kingdom

LIFE EXPECTATION
13-15
YEARS

Market price:
600.00 USD–1,200.00 USD (2023)

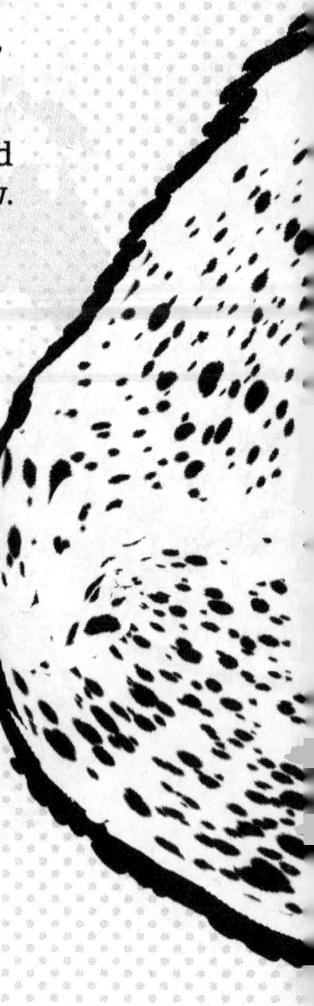

ACF	ACFA	CAA	CCA-AFC	CCC	CCCA	CFA	FB	FIFe	GCCF	LOOF
✓			✓	✓	✓	✓				✓

NZCF	SACC	TICA	WCF
✓	✓	✓	✓

Selection of the most important cat associations in the world: Australia, Canada, China, France, Germany, New Zealand, South Africa, UK and USA

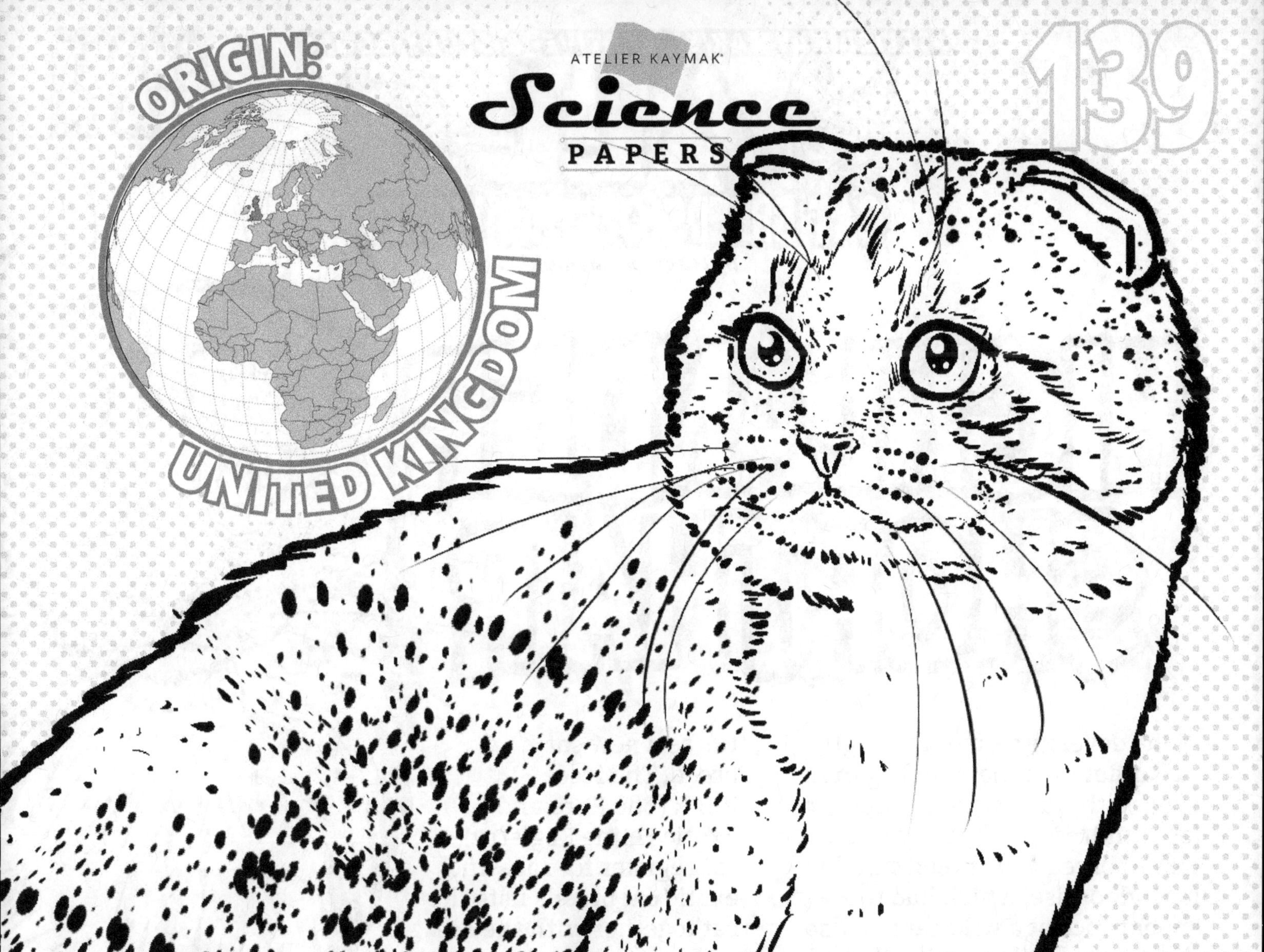
ATELIER KAYMAK®
Science
PAPERS
139
ORIGIN:
UNITED KINGDOM
Recognized since 1986

HIGHLANDER SHORTHAIR

The development of the Highlander Shorthair cat began in 1993 with the American breeder Joe Childress. He had the intention to breed a cat breed suitable for indoor use, which had the appearance of a wild cat but also the personality of a friendly and sociable domestic cat. The result was a Highlander Lynx Bobcat, i.e., a short tail cat with a Jungle Curl cat, which is a hybrid of a reed cat or marshland cat - and an American folding ear cat. Did you get it? Roger. Thus, it took until the year 2008 until the Highlander Shorthair cat was awarded the status as a new cat breed. The free REFR (The Rare and Exotic Feline Registry) cat register for rare and exotic cat breeds lists the cat together with the desert lynx, alpine lynx, and Mojave Bob as part of the broader desert lynx breeding - which despite its name has no lynx genes - and whose name was therefore changed to Highlander in 2005.

Breeding start
1993
United States

ACF	ACFA	CAA	CCA-AFC	CCC	CCCA	CFA	FB	FIFe	GCCF	LOOF

NZCF	SACC	TICA	WCF
		✓	

Selection of the most important cat associations in the world: Australia, Canada, China, France, Germany, New Zealand, South Africa, UK and USA

ORIGIN: UNITED STATES

Recognized since 2008

Market price:
600.00 USD–
1,200.00 USD
(2023)

LIFE EXPECTATION
10-15
YEARS

The short-haired Highlander cat is said to resemble a mini puma, with the folding ears less rolled up than in American or Scottish folding ear cats and the ears - instead of being directed directly forwards - diagonally arranged away from the eyes.

HIGHLANDER

The impressive Highlander cat - which one simply calls Highlander- has a friendly nature and is very sociable. The active, muscular cat brings a lot to the mat and can weigh up to 10 kg. She has a short stump tail, as all Bobtail cats do. Occasionally a genetic defect causes the Highlander cat to grow an extra toe on one or more paws. Polydactyl, multi-fingered like Clipper cat cats and Maine Coons - is not uncommon among Highlander cats. The Highlander cat reminds strongly of its big cousin, the lynx, and looks very similar to the desert lynx. However, it has much smaller and wavy ears, which are a little further apart. Highlander cats have either tabby patterns with Lynx or solid dots. The good-natured cat has an impressive line of ancestors that is easy to lose track of: it is the late generation of a cross between a Desert Lynx cat (a hybrid of lynx and domestic cat) and a Jungle Curl cat (a cross between a Jungle cat - which emerges from the cross between a Bengal cat and a Chausie cat - and last but not least an American Curl cat). That wasn't hard now, was it?

ACF	ACFA	CAA	CCA-AFC	CCC	CCCA	CFA	FB	FIFe	GCCF	LOOF

NZCF	SACC	TICA	WCF
		✓	

Selection of the most important cat associations in the world: Australia, Canada, China, France, Germany, New Zealand, South Africa, UK and USA

ATELIER KAYMAK®

Science

PAPERS

Breeding start
1993
United States

Recognized since 2008

Market price:
600.00 USD–
1,200.00 USD
(2023)

BIG CAT

LIFE EXPECTATION

ISLE OF MAN
SHORTHAIR

The Isle of Man Shorthair resembles the more muscular British Shorthair and is actually a Manx cat with a normal sized tail. It has the same characteristics as a Cymric cat, but without its typical genetic abnormalities. The cat originates, as the name suggests, from the Isle of Man. The British island belongs to the private property of the English Crown and is situated in the middle of the Irish Sea between Ireland and England. Contrary to all expectations, the cat is not bred on the island of the same name, but in New Zealand. The former hereditary diseases of the island cat are systematically eradicated by crossbreeding with healthy domestic cats. That is why the Isle of Man cats look so robust: They are in perfect health and rarely get sick. So far, this young breed has only been recognised by one association: the New Zealand NZCF.

ACF	ACFA	CAA	CCA-AFC	CCC	CCCA	CFA	FB	FIFe	GCCF	LOOF

NZCF	SACC	TICA	WCF
✓			

Selection of the most important cat associations in the world: Australia, Canada, China, France, Germany, New Zealand, South Africa, UK and USA

Breeding start
2010
New Zealand

Recognized since 2013

Market price:
500.00 USD–
800.00 USD
(2023)

LIFE EXPECTATION
8-14
YEARS

Weight

ISLE OF MAN LONGHAIR

The muscular Isle of Man Longhair cat is no less popular than the shorthair version. The ancestors of the playful island cat were brought from the Isle of Man to New Zealand by ship hundreds of years ago. The cat has been moved from one island to the next, so to speak. With the difference that today it is at home on two islands with two climate zones at the other end of the world. The clever and skilful island cat can be loyal like a dog if it likes its owner. She is social, has a friendly attitude and is very playful. This makes her an ideal family animal that also feels at home in a larger community and gets along well with other pets.

Breeding start
2010
New Zealand

LIFE EXPECTATION
8-14
YEARS

Recognized since
2013

Market price:
500.00 USD–
800.00 USD
(2023)

ACF	ACFA	CAA	CCA-AFC	CCC	CCCA	CFA	FB	FIFe	GCCF	LOOF

NZCF	SACC	TICA	WCF
✓			

Selection of the most important cat associations in the world: Australia, Canada, China, France, Germany, New Zealand, South Africa, UK and USA

Weight
7-11
lbs

ORIGIN:
UK, NEW ZEALAND

ANCIENT ORIGIN

Breeding start
1940s
Japan, United States

JAPANESE BOBTAIL SHORTHAIR

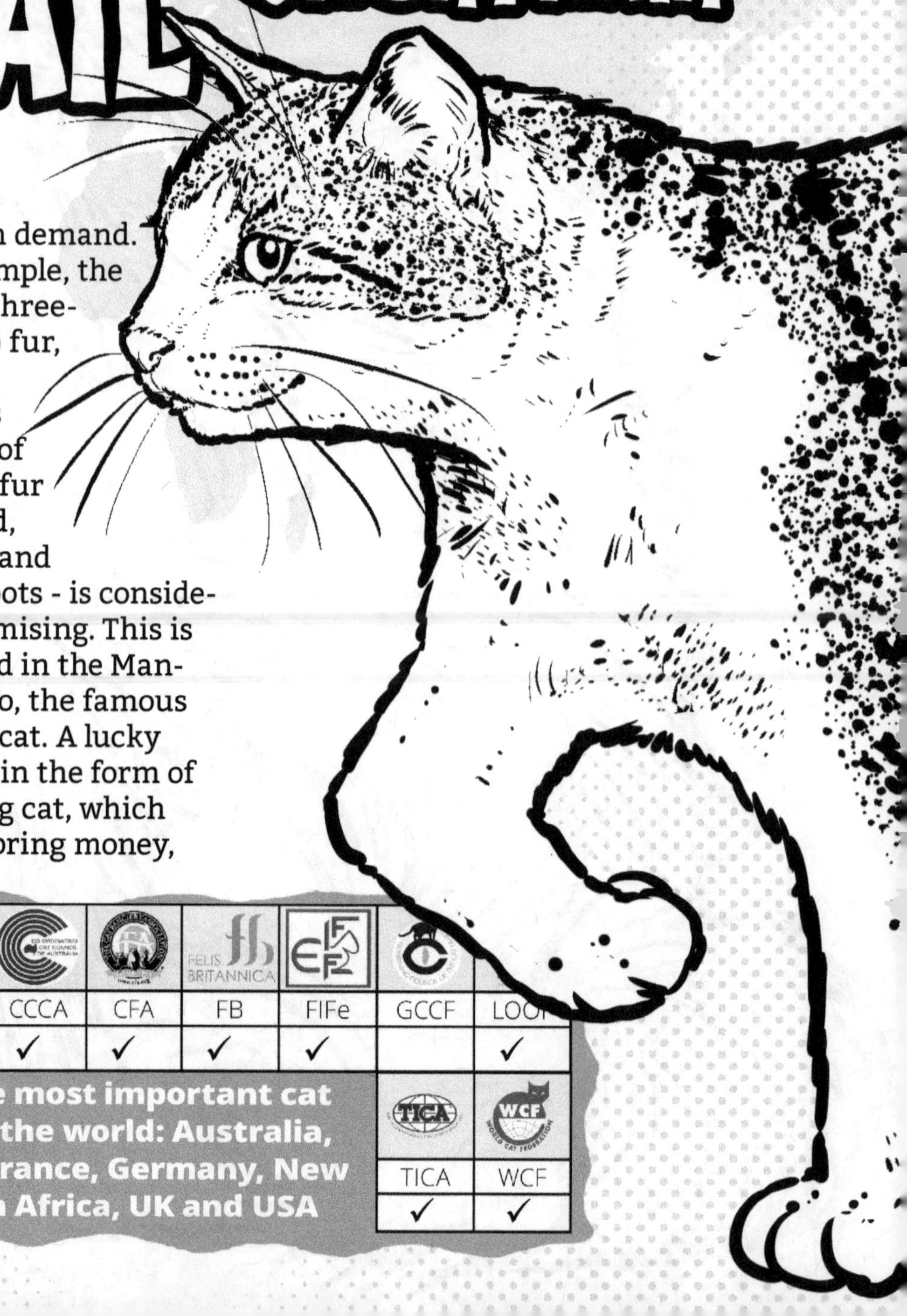

The Japanese Bobtail cat is a natural breed of cat that has lived in various regions of Southeast Asia and China. It arrived in Japan more than a thousand years ago, where the Japanese nobility kept it as a pet. The stumpy tail cat moved into Japanese everyday life and culture. The cat even played a prominent role in art and folklore. In Japan, cats with particular fur drawings are very much in demand. For example, the Mi-Ke (three-colored) fur, which consists mainly of a white fur with red, golden, and black spots - is considered promising. This is reflected in the Maneki-Neko, the famous waving cat. A lucky bringer in the form of a waving cat, which should bring money,

ACF	ACFA	CAA	CCA-AFC	CCC	CCCA	CFA	FB	FIFe	GCCF	LOO
✓	✓	✓	✓		✓	✓	✓	✓		✓

NZCF	SACC	TICA	WCF
✓	✓	✓	✓

Selection of the most important cat associations in the world: Australia, Canada, China, France, Germany, New Zealand, South Africa, UK and USA

prosperity, and luck. Many mistakenly believe that the widespread Chinese version - the waving Golden cat - is a Chinese lucky bringer, which is not true. It is the stylized Japanese Bobtail cat that stands in the entrances of Japan, China, Taiwan, and Thailand and attracts customers.

Recognized since 1991

The most famous folk tale is as follows: A poor grocer picks up a starving stray cat and gives it to eat from the few he has. Out of gratitude, the cat sits down next to the entrance and waves the people into the shop to buy something from the grocer. So, the Maneki-Neko is the lucky charm for the little man. The Japanese stumpy tail cat is slim and muscular. She has slightly larger hind legs than front legs and draws attention to her pom-pom tail with her forward-leaning posture. After World War II, Judy Crawford gave this stumpy tail cat an enormous reputation outside Japan. The American, who had lived in Japan for many years and bred stumpy tail cats during that time, returned to the States in 1968 with a retinue of 38 Japanese Bobtail cats. The cats became a big hit the following year.

Market price:
600.00 USD–1,600.00 USD (2023)

LIFE EXPECTATION

JAPANESE BOBTAIL
LONGHAIR

The long-haired Japanese stub-tailed cat has been known for as long as her short-haired sister, which roamed the streets of the northern provinces of Japan. Despite her elegant appearance, she has been around Japan for centuries as a typical street cat. Of course, one or another cat was kept as a pet, but in general, it did not have a high status in Japanese society. It only gained respect when the silkworms on the silkworm plantations were regularly attacked by mice, rats, and birds, threatening Japan's silk industry. Then the emperor himself intervened on behalf of the cats and practically elevated them to the status of nobility: by imperial decree, it was now forbidden, under severe penalty, to keep the cats from their essential work in the fight against the pests. They were needed in the wild, and no one was allowed to keep these valuable animals as domestic cats anymore. The more recent breeding history began only in 1954 when Japanese cat breeders considered them their national heritage and began to breed them. In the Western Hemisphere, breeding started in the 1960s. In the 1960s, U.S. soldiers stationed in Japan saw Japanese stub-tailed cats of which they took

ACF	ACFA	CAA	CCA-AFC	CCC	CCCA	CFA	FB	FIFe	GCCF	LOOF
✓	✓	✓	✓		✓	✓	✓	✓		✓

NZCF	SACC	TICA	WCF
✓	✓	✓	✓

Selection of the most important cat associations in the world: Australia, Canada, China, France, Germany, New Zealand, South Africa, UK and USA

photos and which then circulated in the United States. Breeder Elizabeth Freret then had three Japanese stub-tailed cats sent to her in 1968, with which she began a breeding program. Although they look alike from a distance, the tail of each Japanese stub-tailed cat is as unique as a fingerprint. Genetically, they are heterozygous cats: these do not suffer from the skeletal abnormalities of other stub-tailed cats, such as Manx cats. The American TICA accepted the long-haired Japanese stub-tailed cat as a separate breed in 1991.

Recognized since 1976

LIFE EXPECTATION 15-18 YEARS

Breeding start 1954 Japan

BOBTAIL CAT

Market price:
600.00 USD

KANAANI

The Kanaani cat is said to have been bred in the early 1990s in the Jerusalem Ha Jeruschalmi Cattery. The breeder, the Russian- Israeli sculptor Doris Polachek, had created the Kanaani cat from an experiment. The breeding succeeded and was continued in the 2000s in Germany. There the Kanaani cat gained great fame because it was hyped together with the Savannah cat and Ocicat cat to the level of a rare and expensive luxury cat. Later the breeding was also continued in the USA. The Kanaani cat is a hybrid cat, which first originated from the crossing of an African wild cat (Felis lybica) with a domestic cat, and in its descendants, Bengal cats and Oriental Shorthair cats were bred gradually. Since 2010 outcrosses with Bengal and Oriental Shorthair cats are no longer allowed, and only direct mating between Kanaani cats is permitted. The Kanaani cat was registered in the year 2000 only by the WCF resident in Germany.

Breeding start 1990 Israel

HYBRID

LUXUS

LIFE EXPECTATION 12-15 YEARS

ACF	ACFA	CAA	CCA-AFC	CCC	CCCA	CFA	FB	FIFe	GCCF	LOOF

NZCF	SACC	TICA	WCF
			✓

Selection of the most important cat associations in the world: Australia, Canada, China, France, Germany, New Zealand, South Africa, UK and USA

Market price:
800.00 USD–
1,500.00 USD

Breeding start
1990s
Russia

KARELIAN BOBTAIL

SHORTHAIR

BOBTAIL CAT

The Karelian Bobtail cat is native to Karelia, where it lives on the coast and on the islands of Lake Ladoga. The Republic of Karelia consists of 25% water (66,000 lakes) and has 49% of forest areas - parts of untouched primeval forests - which are a paradise for Karelian cats. Lake Ladoga, formed by meltwater only 15,000 years ago, is the largest lake in Europe. It extends over a length of 220 km and is 80-120 km wide. On the 500 islands of the lake, a vital vegetation and animal world could develop, which can live largely undisturbed. The Karelian Bobtail cat is a naturally occurring mutation. As is the case with the Japanese Bobtail cat, the Karelian Bobtail cat is not a Manx mutation. This means that the hereditary diseases and health problems associated with Manx syndrome are eliminated - and the cats are born healthy. In contrast to her cousin on the Kuril Islands, the gene of the Karelian Bobtail cat that triggers the stump tail mutation is recessive. That means both parents have to wear it to have puppies with stub tails. Probably this is the reason for the rarity of this cat outside Karelia. The Karelian Bobtail cat was recognized in 1992 in Russia and 1994 by the WCF as a new cat breed.

ACF	ACFA	CAA	CCA-AFC	CCC	CCCA	CFA	FB	FIFe	GCCF	LO

NZCF	SACC	TICA	WCF
			✓

Selection of the most important cat associations in the world: Australia, Canada, China, France, Germany, New Zealand, South Africa, UK and USA

ORIGIN:
ASIA, RUSSIA
ATELIER KAYMAK
Science
PAPERS
Market price:
400.00 USD–
800.00 USD
(2023)
Recognized since
1994
LIFE EXPECTATION
15-20
YEARS
BIG CAT
Weight
10-15
lbs

LONG HAIR

BIG CAT

KARELIAN BOBTAIL LONGHAIR

The hallmark of the Karelian Bobtail cat is its short, kinked tail in the length of 3-14 cm, which looks like a hare's tail. The peculiarity here is that this is individually pronounced with each cat and therefore offers itself like a fingerprint for identification. It is assumed that the Karelian Bobtail contains genes of the Norwegian Forest Cat and the Siberian cat. Among the inhabitants of the Republic of Karelia, the stumpy tail cat is considered a lucky charm. The Karelian Stub-tailed cat has a strong, muscular build, a medium-length dense coat that feels like silk, and a wedge-shaped head. A thick layer of undercoat protects the cat against the Karelian cold, which is for almost all year round with an average of 3 degrees below the refrigerator temperature. The color of its coat can be white, black, blue, red, cream or tortoiseshell, except chestnut brown, point markings, and cinnamon. The fur can also appear in different dilutions and in combination with silver or gold, smoke colors, shaded, and chinchilla. Like other Bobtail cats, she is an amiable and curious cat that quickly befriends humans or other pets.

Breeding start
1990s
Russia

ACF	ACFA	CAA	CCA-AFC	CCC	CCCA	CFA	FB	FIFe	GCCF	LOOF

NZCF	SACC	TICA	WCF
			✓

Selection of the most important cat associations in the world: Australia, Canada, China, France, Germany, New Zealand, South Africa, UK and USA

ATELIER KAYMAK
Science
PAPERS
Recognized since 1994
ORIGIN:
ASIA, RUSSIA
BOBTAIL CAT
Market price:
400.00 USD–
800.00 USD
(2023)
Weight
10-15
lbs
LIFE
EXPECTATION

SHORT HAIR

KHAO MANEE

The Khao Manee (Thai: white jewel) has a second name beside the old Siamese name, which describes it more accurately: Khao Plort (Thai: all white). The Khao Manee cat belongs to an ancient cat breed that lived in the Siamese kingdom of Ayutthaya, which lasted over 400 years (1351-1767). The then Siamese royal house was famous for its beautiful cats, which were strictly guarded, to keep it that way. The breeding standards were meticulously written down in the Tamra Maew (treatises on cats) - a cat textbook - in the form of rhymes and illustrations. The eight-version book included the complete standard of 17 so-called luck bringing cats. Although the Khao Manee is not included in the list of these 17 cats, it was known at least since the Siamese-Burmese wars. Wall paintings in the most important temple near Wat Arun in Bangkok testify to the existence of the Khao Manee and are mentioned in records from the reign of King Nangklao (1824-1851). During the reign of King Chulalongkorn (1868-1910), at the latest, the Khao Manee cat was generally known. The cat has a silky, short, and pure white coat.
It usually has blue eyes, which can also be golden or different in color, the latter being particularly rare. The Khao Manee cat is known for its historical position in Thailand - but is relatively unknown in the West compared to Siamese and Burmese cats. Perhaps it was due to the fact that cats from Thailand were only allowed to be exported since 1999. In the same year, the breeder Coleen Fraymouth took Khao Manee cats with her back to the USA, with whom she immedia-

Breeding start
2000s
United States

ANCIENT ORIGIN

ACF	ACFA	CAA	CCA-AFC	CCC	CCCA	CFA	FB	FIFe	GCCF	LOOF
	✓	✓								

NZCF	SACC	TICA	WCF
			✓

Selection of the most important cat associations in the world: Australia, Canada, China, France, Germany, New Zealand, South Africa, UK and USA

tely started a breeding program and made this cat known in the western world. In 2009 the Khao Manee was registered with TICA. In 2015 she was awarded championship status.

Market price:
200.00 USD–15,000.00 USD

LIFE EXPECTATION
10–14
YEARS

Recognized since 1882

KORAT

In Thailand today, the Korat cat is called Si-Sawat (Thai: color blue-grey) or Maeo Dok-Lao (Thai: cat of the color of the clouds before a rain shower). In spite of these negatively connotated names, the Korat cat still gets a lot of respect, love, and affection from the people. Because it is a national sanctuary in Thailand and is considered a lucky charm, therefore it has a high value in many ceremonies. As its name suggests, it is associated with rain, prosperity, renewal, and a good harvest. In ancient Siam, processions were therefore held in the dry season to stimulate the monsoon: Korat cats were sprinkled with water and carried around in wicker baskets decorated with flowers. The Korat cat is an old cat breed. It was bred in the Kingdom of Ayutthaya (1351-1767) and is thematized in a collection important for the period: the famous Tamra Maew (Thai: treatises on cats). It is a cat textbook of the Siamese royal house in which all the knowledge about the 17 luck bringing cats, the royal court cats - handwrit-

ACF	ACFA	CAA	CCA-AFC	CCC	CCCA	CFA	FB	FIFe	GCCF	LOOF	NZCF	SACC	TICA	WCF
✓	✓	✓	✓	✓	✓	✓	✓	✓	✓	✓	✓	✓	✓	✓

Selection of the most important cat associations in the world: Australia, Canada, China, France, Germany, New Zealand, South Africa, UK and USA

ten - were rhymed and provided with drawings. The Korat cat is in fifth place and is listed with its old name Dork Lao (Thai: Lao flower). In the Tamra Maew, the Korat is described as follows: The hair is smooth, with roots like clouds and tips like silver. The eyes shine like dewdrops on a lotus leaf. The first exact description of a Korat cat in the western hemisphere can be found in London in 1896. At that time, a Korat cat was allegedly presented to the general public in the Crystal Palace as part of a cat exhibition. Unfortunately, this turned out to be the blue Siamese cat and was disqualified. Then the Korat cat was forgotten for more than half a century. It was not until 1959 that Korat cats returned to the West and have been bred in the USA ever since. Already in 1966, she achieved the championship in the CFA.

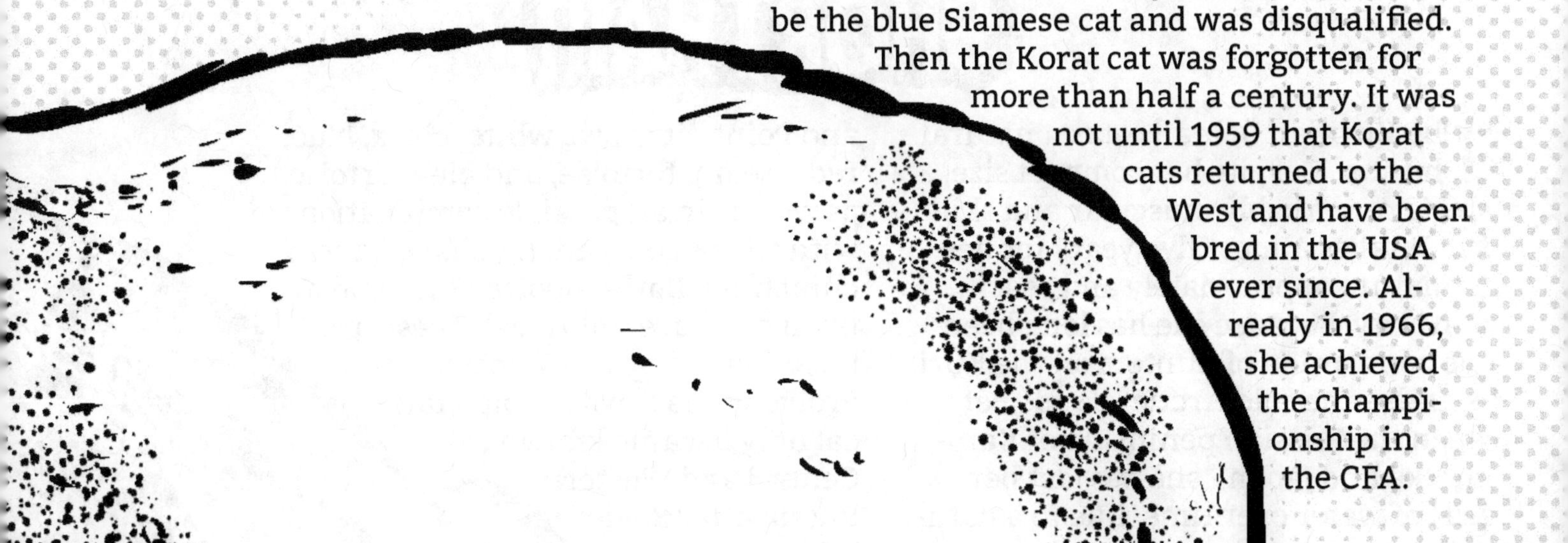

ANCIENT ORIGIN

Market price:
500.00 USD–800.00 USD (2023)

Gangliosidosis

LIFE EXPECTATION
14-16 YEARS

Breeding start

KURILIAN BOBTAIL

SHORTHAIR

The Kurilian Bobtail cat is a natural hunter. Despite her compact size, she is remarkably muscular and skillful. It takes almost five years for her to mature appropriately and reach maturity. Because she has long been the companion of hunters on the Kuril Islands - the cold Arctic islands between the Russian peninsula of Kamchatka and Japan - she has lost her fear of water over time. Due to a naturally occurring dominant mutation, she has a stunted stump tail. But that doesn't bother her much because she is highly mobile and agile. Her stump tail can have between 2 and 10 vertebrae and is characterized by kinks that distort the tail through tight knots. The bushy coat is different for every Kurilian Bobtail cat: except for the colors chocolate-lilac, cinnamon-fawn, and Point - it grows white, black, blue, red, cream, tortoise, and blue tortoise as well as in all possible combinations with silver, gold, smoke, shaded and a chinchilla. Both a spotted coat and a plain color are untypical. These are not accepted by the associations. Probably the Kuril stumpy tail cat only became known in Central and Western Russia after Soviet soldiers and scientists brought it with them from the far-off Siberian coast.

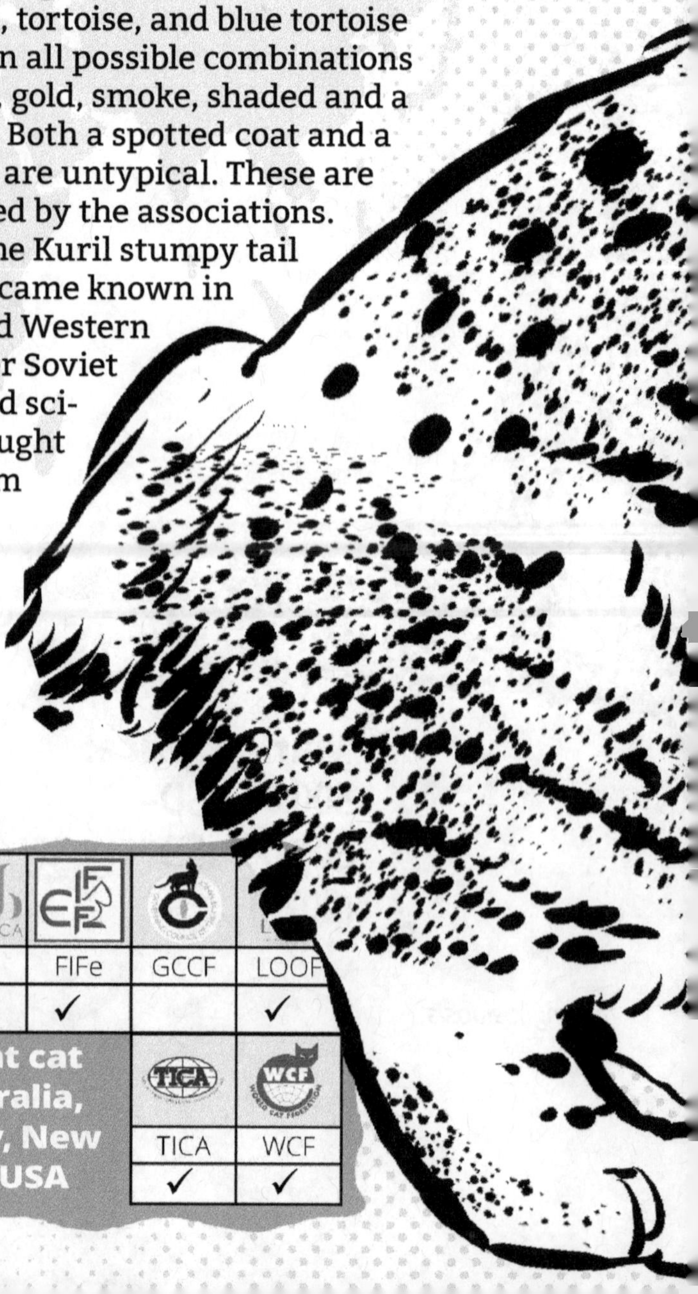

ACF	ACFA	CAA	CCA-AFC	CCC	CCCA	CFA	FB	FIFe	GCCF	LOOF
	✓	✓	✓				✓	✓		✓

NZCF	SACC	TICA	WCF
		✓	✓

Selection of the most important cat associations in the world: Australia, Canada, China, France, Germany, New Zealand, South Africa, UK and USA

ATELIER KAYMAK
Science
PAPERS
163
ORIGIN:
ASIA, RUSSIA
Recognized since 2012
Market price:
700.00 USD–
3,000.00 USD
(2023)
Breeding start
1980
Russia
BOBTAIL
CAT
LIFE EXPECTATION
15-20
YEARS
BIG

Breeding start
1980
Russia

KURILIAN BOBTAIL LONGHAIR

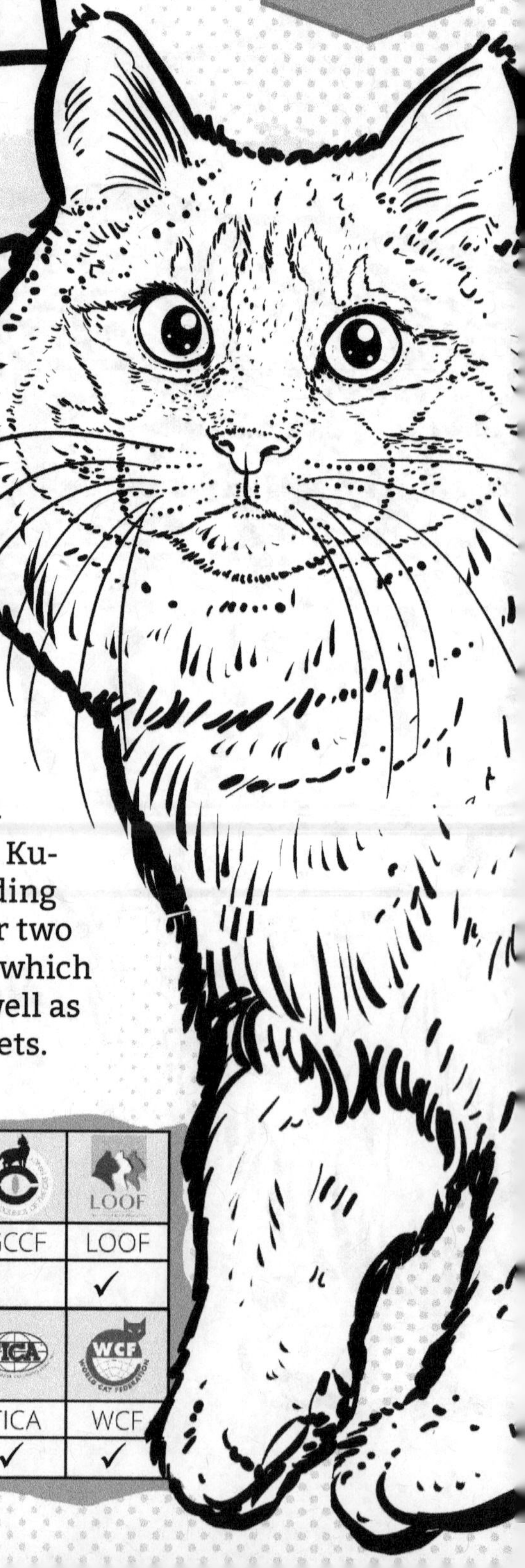

As a resident of the Kuril chain of islands, the Sakhalin Islands, and the Kamchatka peninsula, Kuril's stumpy tail cats are accustomed to a harsh climate. Life on Arctic islands, surrounded by ice-cold water and cutting wind - close to the far eastern coast of Siberia - demands high resistance and robust health from every living creature. Due to its geographical location alone, the Kurile Bobtail cat could be related to the neighboring Japanese Bobtail cat. However, the Kuril Bobtail cat has denser and longer fur than the Japanese cousin. The inhabitants of the Kuril Islands have been breeding stumpy-tailed cats for over two centuries. The lovable cat, which can hunt and fish just as well as humans, is often kept as pets.

ACF	ACFA	CAA	CCA-AFC	CCC	CCCA	CFA	FB	FIFe	GCCF	LOOF
	✓	✓	✓				✓	✓		✓

NZCF	SACC	TICA	WCF
		✓	✓

Selection of the most important cat associations in the world: Australia, Canada, China, France, Germany, New Zealand, South Africa, UK and USA

ATELIER KAYMAK

Science PAPERS

ORIGIN:

ASIA, RUSSIA

BIG CAT

LIFE EXPECTATION 15-20 YEARS

BOBTAIL CAT

Recognized since 2012

Market price:

Breeding start
1950s
United States

LAMBKIN
ALSO NANUS REX

The Lambkin cat - also known as Nanus Rex (nanus, Latin: dwarf) - is one of the rarest cats in the world. There are hardly any breeders and you have to search for a long time and put a lot of money on the table to get a Lambkin cat. It is the result of a cross between two mutated cats: a curly Selkirk Rex and a short-legged Munchkin. In short: a curly Selkirk Rex cat with Lilliputian legs. The Lambkin was developed in 1993 by Terri Harris, an American breeder who had previously bred Munchkin and Napoleon cats. The soft, curly coat - which can grow in all the usual colours - is compared to that of a lamb. The little curls in the fur were decisive for the name of the cat. Fun Fact: Interestingly enough, the curls with which the Lambkin cat was born begin to stretch and smooth again after about 16 weeks - only to contract again after eight to ten months and roll up again into curls. But a rather strange whim of nature. As if the genes could not decide in which direction the journey should go.
The Lambkin cat has a round head with beautiful, ball round eyes - which, when they look into one's eyes with their languishing

LIFE EXPECTATION 14-16 YEARS

REX CAT

Recognized since 1882

ACF	ACFA	CAA	CCA-AFC	CCC	CCCA	CFA	FB	FIFe	GCCF	LOOF
✓	✓	✓				✓	✓	✓	✓	✓

NZCF	SACC	TICA	WCF
	✓	✓	✓

Selection of the most important cat associations in the world: Australia, Canada, China, France, Germany, New Zealand, South Africa, UK and USA

Weight
5-9
lbs

gaze, are really enchanting. As for the Lambkin's build, the stocky hind legs are longer than the front legs, which makes the posture look as if the kitten is about to sprint. The Lambkin cat was accepted by the TICA as a dwarf cat. In countries like Germany, the Animal Protection Act prohibits the breeding of the Lambkin cat.

Dwarf
Cat

Market price:
500.00 USD–
1,000.00 USD
(2023)

LAPERM SHORTHAIR

The story about the origin of the LaPerm cat begins in 1982 - when Linda Koehl found a strange surprise on her farm. One of Linda's farm cats gave birth to a litter of six animals. Among them was an unfortunate, bald kitten, who after six weeks still got a brown curly tabby coat. The curls were a combination of different hair types. They consisted of under- and over-wool and gave the coat a unique texture. The kitten, Curly, was a spontaneous mutation. Mrs. Koehl was not a breeder and had no idea what she was looking at. She had nothing to do with cats and mutations before. Curly was to become the ancestral mother of all LaPerm cats following her due to the permanent mutation. Curly grew up, became a pretty queen, and, following her genetic order, soon became pregnant (Mrs. Koehl let the cats do what they wanted outdoors until then). Curly now got five cute males, who were born as naked as she had been herself. Currently, Linda's interest was aroused. She decided to take the cats into care and started a breeding program. She crossed Curly's Offspring with normal domestic cats and increased the number of cats. Before she presented her lured cats to the public, she chose a suitable name that would do justice to the animals: LaPerm - the permanent wave. The cat gained so many followers over time that in 1997 the LaPerm Society of America was founded by a number of interested breeders. In 2003, TICA awarded the breed championship status, followed by the GCCF in 2012.

Breeding start
1997
United States

ACF	ACFA	CAA	CCA-AFC	CCC	CCCA	CFA	FB	FIFe	GCCF	LOOF
✓	✓	✓				✓	✓	✓	✓	✓

NZCF	SACC	TICA	WCF
✓	✓	✓	✓

Selection of the most important cat associations in the world: Australia, Canada, China, France, Germany, New Zealand, South Africa, UK and USA

ORIGIN:
UNITED STATES
Recognized since 2003
LIFE EXPECTATION
10-15 YEARS
Market price:
100.00 USD–
1,500.00 USD
(2023)
Weight

LAPERM LONGHAIR

The LaPerm cat is often born bald. It has a tabby pattern that can look mackerel, tabby, spotted, or ticked and seems as if it has been tattooed into the skin. The etymology of the LaPerm cat's name is as interesting as the unique cat itself: the article La derives from a French naming convention that has been adopted from the Chinook Indian tribe into their Chinook Wawa - their language - when new words are accompanied by a particular article. The term Perm is derived from the description of the Rex mutation and comes from a permanent wave, which gives the LaPerm cat a shaggy coat and a distinctive, bushy tail similar to that of a fox. The full name the Dalles LaPerm denotes the origin of the unique LaPerm cat: the Dalles Farm in Oregon.

Breeding start
1997
United States

ACF	ACFA	CAA	CCA-AFC	CCC	CCCA	CFA	FB	FIFe	GCCF	LOOF
✓	✓	✓				✓	✓	✓	✓	✓

NZCF	SACC	TICA	WCF
✓	✓	✓	✓

Selection of the most important cat associations in the world: Australia, Canada, China, France, Germany, New Zealand, South Africa, UK and USA

ATELIER KAYMAK®

Science

PAPERS

Recognized since **2003**

Market price:
100.00 USD–
1,500.00 USD
(2023)

LIFE EXPECTATION
10-15
YEARS

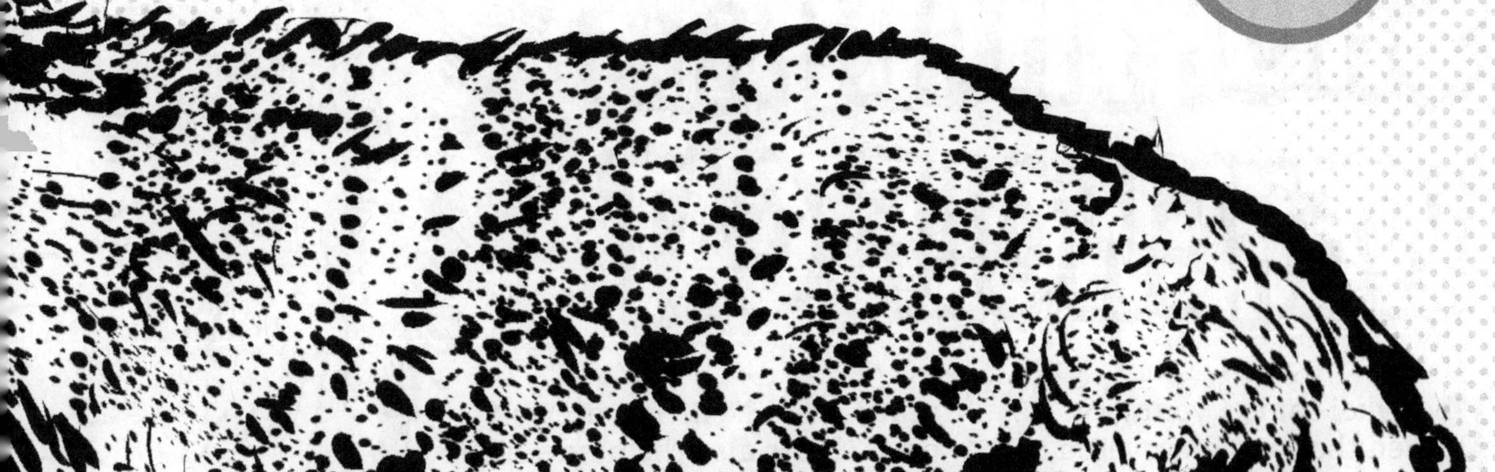

LYKOI SHORTHAIR ALIAS WEREWOLF CAT

The Lykoi cat is a young cat breed that was discovered in 2010 in Tennessee, USA. Patti Thomas found two unusual-looking cats in a shelter in Virginia. The poor cats lacked the complete undercoat, so they looked ripped and, in some places, completely bald. The effect was intensified by the interesting eye color, which is otherwise known by larger predators. The cats looked really scary, just like little werewolf cats. This is also the name of the Lykoi cat: Lykoi means wolves in Greek. One year later, in 2011 - again in Tennessee- Johnny and Britney Gobble noticed two more cats with the same hair mutation. They had the cats examined by a veterinarian to make sure that the hairlessness of the cats was not the result of a disease.

Breeding start
2010s
United States

ACF	ACFA	CAA	CCA-AFC	CCC	CCCA	CFA	FB	FIFe	GCCF	LOOF
✓	✓	✓	✓	✓	✓	✓	✓	✓	✓	✓

NZCF	SACC	TICA	WCF
✓	✓	✓	✓

Selection of the most important cat associations in the world: Australia, Canada, China, France, Germany, New Zealand, South Africa, UK and USA

After thorough investigation, he found that the mutation was of natural origin and was caused by a gene defect. However, he was unable to establish a connection with the Sphynx or Devon gene. The Lykoi has been a TICA-recognized breed since 2017.

In countries like Germany, the Animal Welfare Act prohibits the breeding of the Lykoi cat.

LIFE EXPECTATION
12-15
YEARS

Recognized since
2017

Market price:
1,800.00 USD–
3,000.00 USD
(2025)

LUXUS

LYKOI LONGHAIR ALIAS WEREWOLF CAT

The Lykoi cat is the only known cat with the coat coloring roan - a mixture of white and other colors, which does not lighten or change with age. The Lykoi cat is also known to lose all its hair as it grows older, but it will grow back again after a few months. The long-haired coat gives her an even wilder look. Because she doesn't have the undercoat, she doesn't like the cold. One of the first breeders of the Lykoi cat, veterinarian Dr. Johnny Gobble, describes her as a very loyal animal. She is very curious about her surroundings and has a unique talent for scents. In direct sunlight, the exposed, pink skin of the Lykoi cat turns darker. The coat of the Lykoi cat is probably unique among cats: the continuous black, reddish-grey fur resembles very much that of an opossum. The Lykoi cat has a wedge-shaped head and a supple body. In countries like Germany, the animal protection law prohibits the breeding of the Lykoi cat.

Breeding start
2010s
United States

ACF	ACFA	CAA	CCA-AFC	CCC	CCCA	CFA	FB	FIFe	GCCF	LOOF
✓	✓	✓	✓	✓	✓	✓	✓	✓	✓	✓

NZCF	SACC	TICA	WCF
✓	✓	✓	✓

Selection of the most important cat associations in the world: Australia, Canada, China, France, Germany, New Zealand, South Africa, UK and USA

ATELIER KAYMAK

Science

PAPERS

ORIGIN:

UNITED STATES

LUXUS

LIFE EXPECTATION
12-15 YEARS

Recognized since 2017

Market price:
1,800.00 USD–
3,000.00 USD
(2025)

MAINE COON

ALSO AMERICAN

Breeding start
1960s
United States

As a true giant among domestic cats - it can measure 1.20 meters from the tip of its nose to the end of its tail and weigh up to 12 kg - the Maine Coon cat is North America's answer to the Norwegian Forest Cat and the Siberian Cat. However, the Maine Coon has a wilder appearance thanks to its lynx-like face. It has large, long ears from which tufts of hair grow at the end of which brush hairs often - but not always - grow as with lynxes. In combination with widely spaced, penetrating eyes and a relatively long face - the cat looks even bigger. The overall body, the luxuriant, fluffy Longhair fur, broad paws, and the mighty, big tail, which looks like a giant feather duster - let the cat appear gigantic.

ACF	ACFA	CAA	CCA-AFC	CCC	CCCA	CFA	FB	FIFe	GCCF	LOOF
✓	✓	✓	✓	✓	✓	✓	✓	✓	✓	✓

NZCF	SACC
✓	✓

Selection of the most important cat associations in the world: Australia, Canada, China, France, Germany, New Zealand, South Africa, UK and USA

TICA	WCF
✓	✓

TOP 10

ATELIER KAYMAK

Market price:
1,000.00 USD–
3,000.00 USD
(2025)

LUXUS

Recognized since 1976

LIFE EXPECTATION
12-15 YEARS

ANCIENT ORIGIN

Hypertrophic cardiomyopathy (HCM), Spinale Muskelatrophien (SMA), hip joint diseases, (with pure white coat) Congenital hearing loss, deafness, disturbances of the sense of equilibrium, Eye tremor

BIG

The cat breed, as its name suggests, is native to Maine - the most northeastern state in the USA, bordering Canada. There are several explanatory models for the second part of the name.

The first story about the origin of the Maine Coon cat has its roots in the English word Racoon and is based on the assumption that the Maine Coon cat was created by mating a domestic cat with a raccoon. The comparison is evident because the Maine Coon cat likes water just like a raccoon, is not afraid of it, and even plays in it. Then she's almost as skilled. Because the toes of the Maine Coon cat are far apart, she can literally take things in her hand: before eating, she picks up her food with her paw and, like a raccoon, dips it in water beforehand. She can also easily open doors, turn taps, and retrieve things like a dog - which is why the Maine Coon cat has been given the nickname Dog cat.

According to another legend, the Maine Coon cat is a descendant of Captain Coon's long-haired ship cats, who jumped off the board every time the merchant ship docked to have fun with harbor cats. Whenever long-haired cats were born, the settlers shouted, Another Coon!

The third story is based on the assumption that the Maine Coon cat is a descendant of Marie-Antoinette's Longhair cats. The unpopular Queen of France and Navarre, whose love of longhaired cats was well known, actually sent some

of her private property on ships to Maine. Perhaps the wise woman's plan was to move before the French Revolution and save her head.

A fourth attempt of explanati-on puts the Maine Coon cat as a descendant of Viking cats, who accompanied the North men on their voyages of discovery to America - and who is said to have mated with native cats on shore leave.

The fifth and probably most credible explanatory model moves the roots of the Maine Coon cat to Great Britain. From there, the British Shorthair cat arrived in the New World as a ship cat during the colonization of New England. At that time, it was an unwritten law that cats had to travel on long ship passages as rat hunters. Otherwise, sailors threatened to mutiny - also because ship cats were considered lucky charms. Even the captains firmly believed in it and refused to cast off without a ship's cat. Many of these good, skillful cats later lived in immigrant settlements. They became so popular that they were crossbred with Turkish Angora cats and bred as race cats.

The foundation of the American CFA - the Cat Fancy Association is owed to the Maine cats, as the Maine Coon cats were called at that time. And they were Maine Coon cats, which were shown in 1871 on the first cat exhibition of the world in the famous Chrystal Palace in London.

MANDALAY

Breeding start
1980s
New Zealand

The Mandalay cat, which was called Australian Bombay until it was renamed in 2013, was inadvertently conceived in New Zealand in the early 1970s. The name is a homage to the origin of the Burmese cat - Myanmar - formerly Burma, which borders Thailand in the east. Mandalay - today the second-largest city in Myanmar with 1.2 million inhabitants - was founded by King Mindon in 1857 and replaced the former capital Amarapura, 11 km to the south, which had just been built by King Bodawpaya in 1783. The Mandalay cat is a cross between an elegant Burma cat and a stray black cat. The litter with the enchanting kittens convinced the two cat lovers, Horton Dorothy and Hogen Pat, immediately so that they took up the breeding of the Bombay Mandalay cats as they called them. In the 1970s and 1980s, they crossed Burmese cats with red farmer cats and Burmese cats with black domestic cats. Shortly afterward, genes of Abyssinian cats were added, because of which eyes were more beautiful - and Siamese cats. The muscular Mandalay cat has splendid, golden to orange eyes and carries a monochrome, silk glossy - primarily black - fur. The temperamental but incredibly peaceful cat is - like her ancestral mother, the Burma cat - very talkative and likes to communicate loudly with her owner. The family-related Mandalay cat is very social towards children and dogs and is therefore excellently suited as a family cat. Because there are not even a handful of breeders - in Australia, there are only four - the cat is very rare and is considered a rarity. Mandalay was recognized as a new breed by the Australian ACF, the French FIFe, the New Zealand CFA, and the German WCF.

ACF	ACFA	CAA	CCA-AFC	CCC	CCCA	CFA	FB	FIFe	GCCF	LOOF
✓								✓		

NZCF	SACC	TICA	WCF
✓			✓

Selection of the most important cat associations in the world: Australia, Canada, China, France, Germany, New Zealand, South Africa, UK and USA

Market price:
400.00 USD–
1,700.00 USD
(2023)

Recognized since 1990

LIFE EXPECTATION 10-17 YEARS

SHORT HAIR

TOP 10

MANX

There are many myths about the origin of the famous tailless cat. One of them takes place in Noah's Ark, where two Manx cats trying to escape to the Ark quarreled about who would be allowed to board through the drop gate first. While the brawlers hissed noisily in the middle of the rising water, Noah closed the doors and accidentally cut off the cocks of the two brawlers. Since then, the Manx cats have no tail anymore. Of course, this is nonsense - the tail was not chopped off in reality but could not unfold through a lousy gene mutation at all. Scientists justify this with the so-called genetic bottleneck: selection in isolation due to extreme inbreeding. The name of the cat comes from the language of the native Celtic islanders of the Isle of Man - a small mountainous island between Ireland and the United Kingdom with about 85,000 inhabitants. Since the Viking invasion, the inhabitants of the small, rocky island have been under the influence of changing rulers over the course of 1000 years: Scandinavians, Scots, and English - until the island finally ended up in the private possession of the Queen of England. So, what does that have to do with cats? In the course of Caesar's conquests, the Egyptian cats brought along by members of the army also

Breeding start
1950s
United States

ACF	ACFA	CAA	CCA-AFC	CCC	CCCA	CFA	FB	FIFe	GCCF	LOOF
✓	✓	✓	✓	✓	✓	✓	✓	✓	✓	✓

NZCF	SACC	TICA	WCF
✓	✓	✓	✓

Selection of the most important cat associations in the world: Australia, Canada, China, France, Germany, New Zealand, South Africa, UK and USA

BOBTAIL CAT

ORIGIN: ISLE OF MAN, EGYPT, MIDDLE EAST

arrived on the small Isle of Man, which offered itself as a highway gas station in the middle of the Irish Sea. During many of these breaks, cats repeatedly landed, remained, and multiplied under the protection of isolation. With many island cats, this led to the aforementioned formation of the breed-typical stumpy tails. On the British islands, the Manx- cat is conducted as its own cat-race since the 1920s.

Market price:
100.00 USD–
1,400.00 USD
(2023)

Recognized since 1882

Manx Syndrom (spina bifida); (with pure white coat) Congenital hearing loss, deafness, disturbances of the sense of equilibrium, Eye tremor

LIFE EXPECTATION
14-16 YEARS

MEKONG BOBTAIL

The historically widespread Mekong Bobtail cat in Southeast and Central Asia - could well be a distant relative of the similar-looking Japanese Bobtail cat. The eponym of the Mekong Bobtail cat, the river Mekong, is the so-called lifeline of Southeast Asia. An approximately 4,500 km long river that rises in the Tibetan highlands in the north and flows into the sea in Vietnam in the south. It flows through six countries: China, Myanmar (formerly Burma), Laos, Thailand, Cambodia, and Vietnam. The Mekong Bobtail cat is a naturally occurring cat in these areas and has long been known as a Thai cat in the population. As early as 1883, Charles Darwin mentioned strange-looking cats in his book The Variation of Animals and Plants under Domestication:

Everywhere in these vast areas, particularly in the Malay archipelagos, Siam, China, and Burma, cats carry short tails that are about half the length one might expect and often end up in a kind of knot.

Recognized since 2004

ACF	ACFA	CAA	CCA-AFC	CCC	CCCA	CFA	FB	FIFe	GCCF	LOOF	NZCF	SACC	TICA	WCF
														✓

Selection of the most important cat associations in the world: Australia, Canada, China, France, Germany, New Zealand, South Africa, UK and USA

In 1884, somehow, presumably by diplomats, a pretty stumpy tail cat arrived from Bangkok to the West. In Russia years later, there was an attempt to breed away the stump tails of Thai cats for whatever reason, but this failed miserably. The cat was then forgotten for more than 50 years. Only in the 1980s, something happened again: returning Russian diplomats had some Thai cats in their luggage on their journey home from Southeast Asia. Sometime later, Thai stub-tailed cats also arrived in Russia from Iran. On the basis of these cats brought into the country, breeders began again to crank up the experimental breeding program with the Thai stub-tailed cats again.

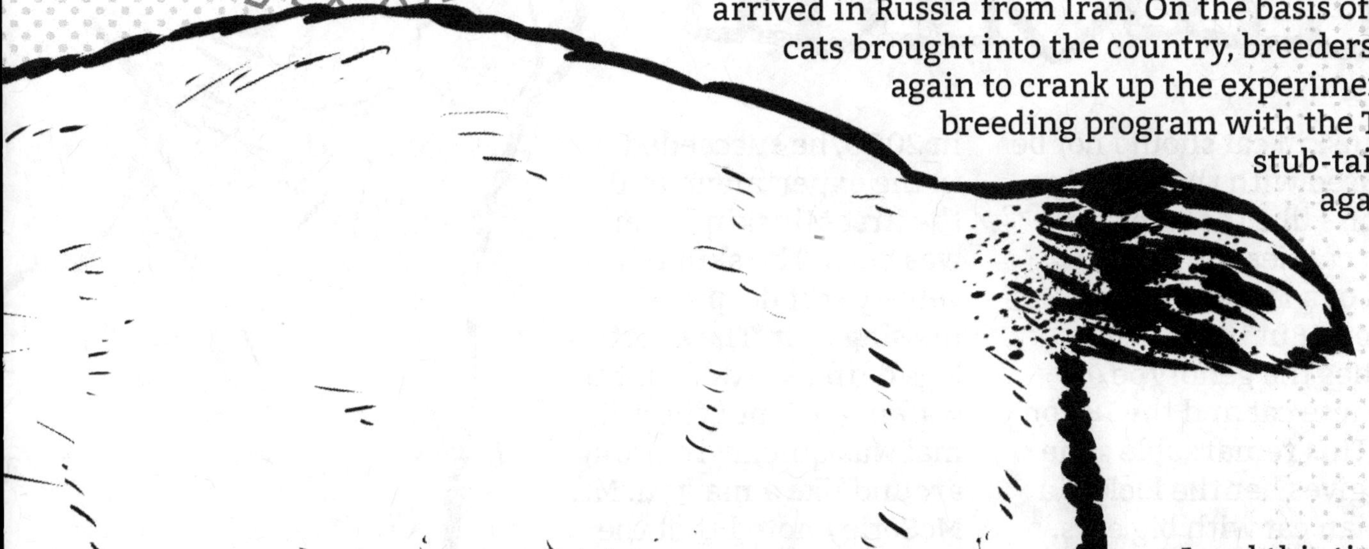

And this time, it worked: the first stumpy tail cat bred in Russia looked very similar to a Siamese cat with a Colorpoint pattern. In order to distinguish it from the original Thai stub-tailed cat, a concise name was chosen that respected the origin of the new breed: the Mekong stub-tailed cat was born. In 2004 it was recognized by the WCF as a new cat breed.

Market price:
700.00 USD–
800.00 USD
(2023)

Breeding start
1980s

LIFE EXPECTATION

MINSKIN

The Minskin cat should not be confused with the Bambino cat, which - like the Minskin cat - also carries the genetic material of a Munchkin and Sphynx cat. But the Minskin cat also has the genotype of the Burmese cat and the Devon Rex cat. This remarkable gene cocktail gives her the look of a noble Asian cat with big ears. The Minskin cat was born in 1998 in Boston, Massachusetts, where the breeder, Paul Richard McSorley, bred a naked cat to a dwarf cat: he crossed a Sphynx with a Munchkin cat. His declared aim was to create a new breed of cat with as little coat as possible and extremely short legs. Two years later, in 2000, he succeeded in the experiment, and the first Minskin kitten was born. The skin felt velvety soft despite the missing hair. The short legs didn't slow down the walking act, but the animal was quickly roaming around like a marten. Mr. McSorley noted that the Minskin gene is inherited recessively in the effect on the sparse hair and dominantly in the expression of the short legs. A heterozygous (mixed) Minskin can therefore also produce long-legged Offspring. The lovable Minskin cat is very self-confident and loves the closeness of humans.

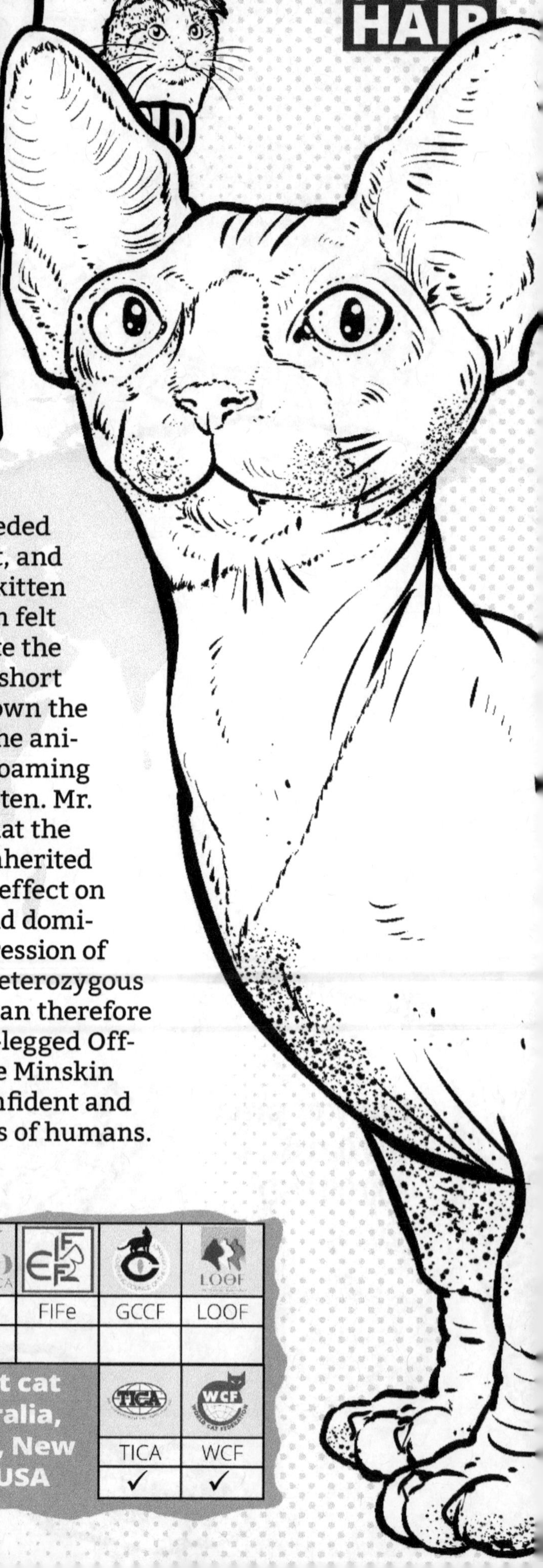

ACF	ACFA	CAA	CCA-AFC	CCC	CCCA	CFA	FB	FIFe	GCCF	LOOF

NZCF	SACC	TICA	WCF
		✓	✓

Selection of the most important cat associations in the world: Australia, Canada, China, France, Germany, New Zealand, South Africa, UK and USA

TICA and WCF have approved the Minskin cat as a temporary, experimental breed. In countries like Germany, the Animal Protection Act prohibits the breeding of the Minskin cat.

Market price:
600.00 USD–
1,700.00 USD
(2023)

Breeding start
1998
United States

HAIRLESS CAT

Dwarf cat

LIFE EXPECTATION
12-15
YEARS

Recognized since
2008

Breeding start
1980s
United States

MUNCHKIN
SHORTH

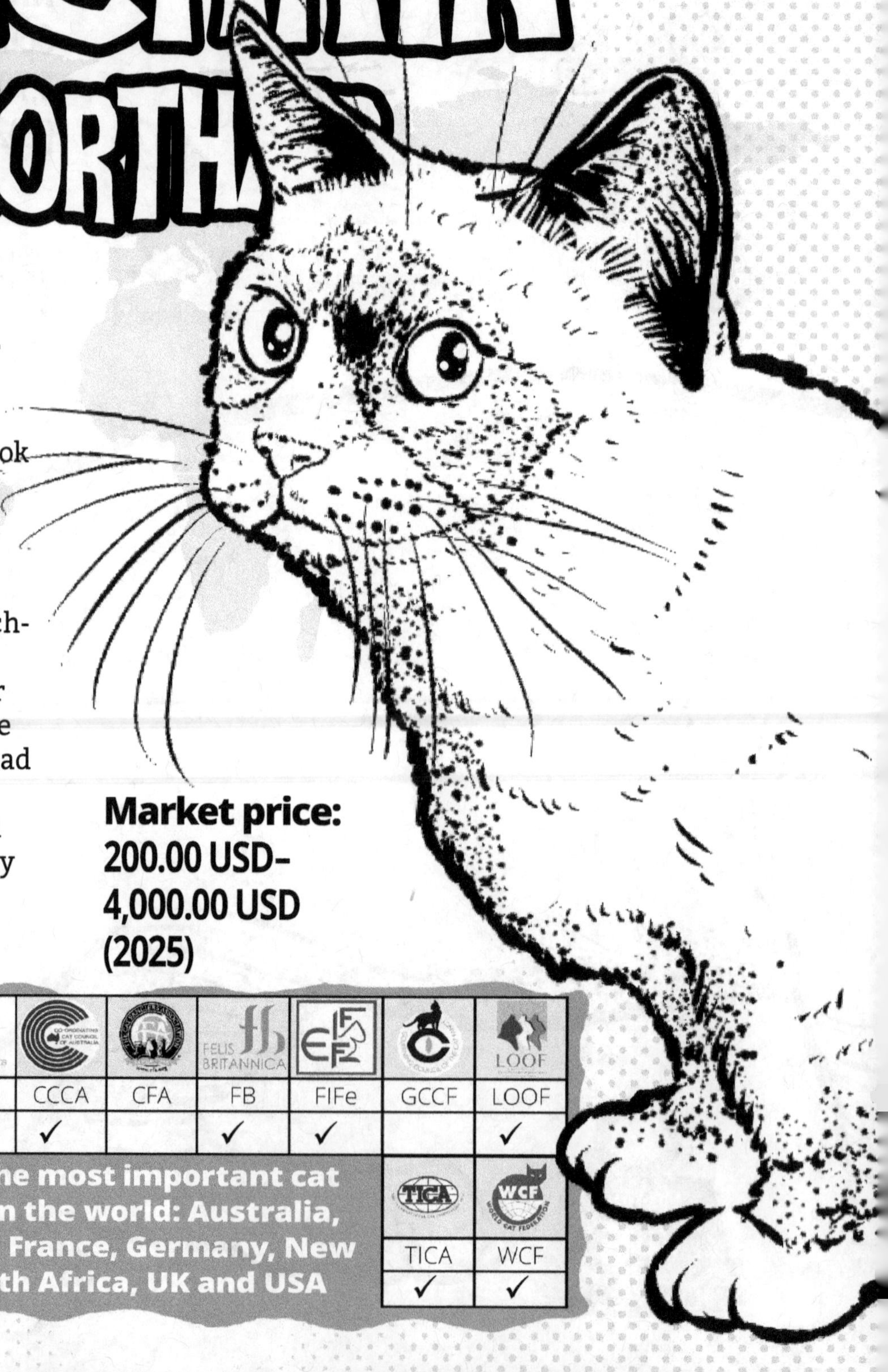

The breeding history of the Munchkin Cat begins in 1983 with a drama: The music teacher Sandra Hochenedel rescued two highly pregnant kittens, which had fled from a biting dog and took one of the kittens with herself. Blackberry, as the cat was now called - very soon had offspring. Blackberry's litter also included short-legged mutations. Mrs. Hochenedel gave one of her kittens, the little tom Cat Toulouse, to her boyfriend, Kay LaFrance. Because Toulouse was not neutered and had regular exercise, countless kittens with short legs soon roamed the neighborhood. Assuming they were alien Cats, Hochenedel and LaFrance contacted Dr. Solveig

Market price:
200.00 USD–
4,000.00 USD
(2025)

ACF	ACFA	CAA	CCA-AFC	CCC	CCCA	CFA	FB	FIFe	GCCF	LOOF
✓					✓		✓	✓		✓

NZCF	SACC	TICA	WCF
	✓	✓	✓

Selection of the most important cat associations in the world: Australia, Canada, China, France, Germany, New Zealand, South Africa, UK and USA

Pflueger, one of the chairmen of TICA. He consulted Dr. David Biller, head of radiology at the Kansas State University School Veterinary Medicine. Dr. Biller found, after a series of investigations, that the Munchkin mutation did not cause any damage to health. There have been - and still are - many controversies about the health risks in breeding Munchkin Cats. Despite Biller's research, animal rights activists are still firmly convinced that the genetic mutation of the Munchkin Cat is associated with health problems and that the animal suffers from them. In countries like Germany, the animal protection law prohibits the breeding of the Munchkin Cat.

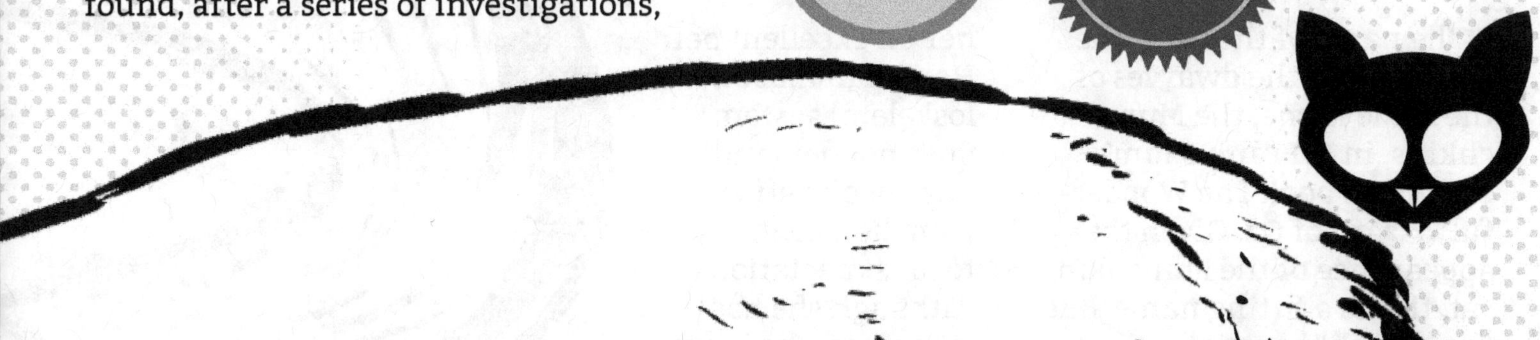

MUNCHKIN LONGHAIR

The name of the cat is the name for the dwarves of the same name, the Munchkins, in L. Frank Baum's children's book The Wonderful Wizard of Oz. Given the appearance of the Munchkin cat, this is a fitting name. But the Munchkin cat also has a second name in the Anglo-Saxon world - and that's Sausage cat. And this one sounds less nice because it's not the cat's fault that it looks like a walking sausage. Completely incomprehensible since the Munchkin cat is a likable and lovable creature. She is intelligent, curious, and playful, and her friendly nature makes her an excellent pet. Her small musculoskeletal system does not demand any special effort from her, contrary to all expectations. Although she has short legs, she is surprisingly agile and whizzes through the apartment like a Fiji arrow. But let's start all over again. Short-legged cats - the forerunners of the Munchkin cat - are said to have been sighted as early as the 1940s. According to an English veterinarian, there have been four generations of short-legged kittens in

ACF	ACFA	CAA	CCA-AFC	CCC	CCCA	CFA	FB	FIFe	GCCF	LOOF
✓					✓		✓	✓		✓

NZCF	SACC	TICA	WCF
✓	✓	✓	✓

Selection of the most important cat associations in the world: Australia, Canada, China, France, Germany, New Zealand, South Africa, UK and USA

ORIGIN:

EUROPE, UNITED STATES

Great Britain - who, apart from their short legs, is said to have lived in perfect health. There were also reports of dwarf cats in 1956 from the Soviet Union - wherein Stalingrad - which was renamed Volgograd in 1961 - mutated cats with short legs are said to have lived. In countries like Germany, the animal protection law prohibits the breeding of the Munchkin cat.

LIFE EXPECTATION 12-14 YEARS

Recognized since 1994

Breeding start 1980s United States

Dwarf Cat

Market price:
200.00 USD–
4,000.00 USD
(2025)

Weight
6-9

NEBELUNG

The name Nebelung is a nomenclature word from the German word Nebel (fog) and the epic poem Nibelungenlied. The young breeding history of the rare Nebelung cat began in the USA in the 1980s. At that time, the magicians and tamers Siegfried & Roy in Las Vegas celebrated huge successes. In breathtaking shows, the successful duo performed their tricks with fabulous big white lions and tigers that thrilled millions of people. Probably the inspiration and motivation for the breeding of this cat breed is here as well as the heroic epic. The Nebelung cat originated from the crossing of a Russian Blue cat and a domestic cat with Angora genes. Two of the longhaired newborns - Siegfried and Brunhilde - were adopted by Cora Cobb, who soon started breeding the Nebelung cat with them. We remember the Nibelungenlied saga with Siegfried the Dragon Slayer and Brunhilde the Valkyrie. The old Germanic and Scandinavian heroic saga was the inspiration for the names. The external appearance of the Nebelung cat resembles the Russian Blue cat - from it differs only in relation to the length of the fur. The Nebelung cat has a very soft coat with a dense undercoat. It has long legs and a long fluffy

ACF	ACFA	CAA	CCA-AFC	CCC	CCCA	CFA	FB	FIFe	GCCF	LOOF
✓					✓		✓	✓		✓

NZCF	SACC	TICA	WCF
✓	✓	✓	✓

Selection of the most important cat associations in the world: Australia, Canada, China, France, Germany, New Zealand, South Africa, UK and USA

Breeding start 1980s United States

ORIGIN:

EUROPE, UNITED STATES

tail. The head with its relatively short hair rests on a rather long neck, from which a lion-like mane grows and merges onto the chest. On the paws, tufts grow out of light down between the toes. The Nebelung cat was recognized by TICA in 1987.

Market price:
350.00 USD–1,800.00 USD (2023)

Recognized since 1987

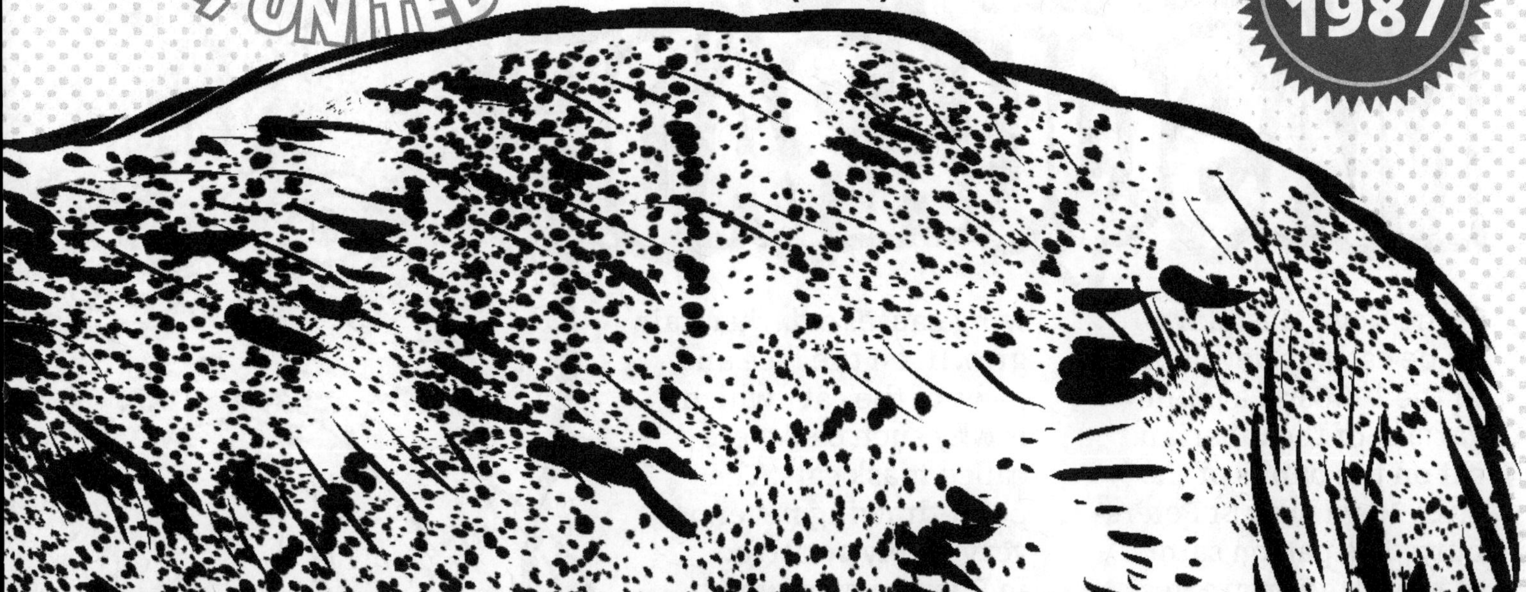

LIFE EXPECTATION 15-18

NEVA MASQUARADE

The Neva Masquarade cat was developed in the 1980s as a point variant of the Siberian cat on the banks of the river Neva near St Petersburg. The patron saint of the name originates from Lake Ladoga 74 km away - the largest lake in Europe - and crosses the megacity before it flows into the Baltic Sea in the west. In Saint Petersburg the breeders were working on the Siberian cat - when the colour variation with the point drawing was created. The point drawing causes a darker hair growth on the face and paws of the cat - which is why such cats were called mask cats. The point-drawing never or very rarely occurred in Russian cats, which made the kitten a half sensation. The mask drawing gives the impression that the cat is wearing a medieval mask - from which ice-blue eyes look out. The appearance of the Neva Masquarade cat - with the exception of the mask drawing - does not differ from the Siberian cat at all.

ACF	ACFA	CAA	CCA-AFC	CCC	CCCA	CFA	FB	FIFe	GCCF	LOOF
✓						✓	✓	✓	✓	✓

NZCF	SACC	TICA	WCF
	✓	✓	✓

Selection of the most important cat associations in the world: Australia, Canada, China, France, Germany, New Zealand, South Africa, UK and USA

While some associations classify the Neva as their own class, others group them with Siberian cats. The Neva Masquarade cat was presented as an official state gift to foreign dignitaries such as the Finnish President and the Japanese Governor of Akita Prefecture.

BIG CAT

Breeding start 1980s Soviet Union

Recognized since 1987

LIFE EXPECTATION 12-15 YEARS

Market price:
700.00 USD–
1,200.00 USD

BIG CAT

NORWEGIAN FOREST CAT

The large Norwegian Forest cat has adapted as a naturally developed cat breed to survival in the cold and wet fjords of Norway. It was probably brought to Scandinavia on Viking longboats, which had previously carried trade out of their region into the world and brought back wild cats from Siberia, Turkey, Central Europe and Great Britain. In the rough climate of Scandinavia, the bushy fur, characteristic for them today, grew as a protection against the weather in the course of time. They had to adapt, for better or for worse, to survive the harsh Norwegian winters. During the Second World War, the Norwegian Forest cat threatened to die out. Because of the strong increase of stray domestic cats the fate threatened the cat to disappear into the former cats and as a cat species. However, this could be prevented thanks to a careful breeding program of the homeland federation for the preservation of the Norwegian forest cat. In 1972 the cat was even officially named Norway's national cat and has since been called Norsk Skogkatt.

Breeding start
1930s
Norway

ACF	ACFA	CAA	CCA-AFC	CCC	CCCA	CFA	FB	FIFe	GCCF	LOOF
✓	✓	✓	✓	✓	✓	✓	✓	✓	✓	✓

NZCF	SACC	TICA	WCF
✓	✓	✓	✓

Selection of the most important cat associations in the world: Australia, Canada, China, France, Germany, New Zealand, South Africa, UK and USA

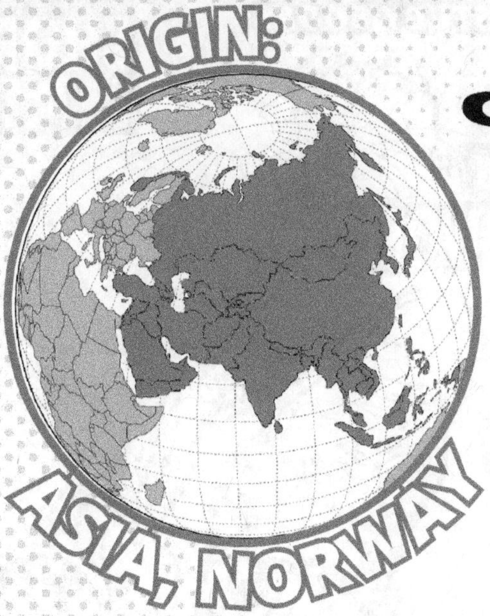

ATELIER KAYMAK®

Science

PAPERS

The robust cat is very popular both in its native Scandinavia and in France.

Recognized since 1970

Market price:
200.00 USD–4,500.00 USD (2025)

LIFE EXPECTATION 14-16 YEARS

Glykogenose IV, hip joint diseases, (with pure white coat) Congenital hearing loss, deafness, disturbances of the sense of equilibrium, Eye tremor

OCICAT

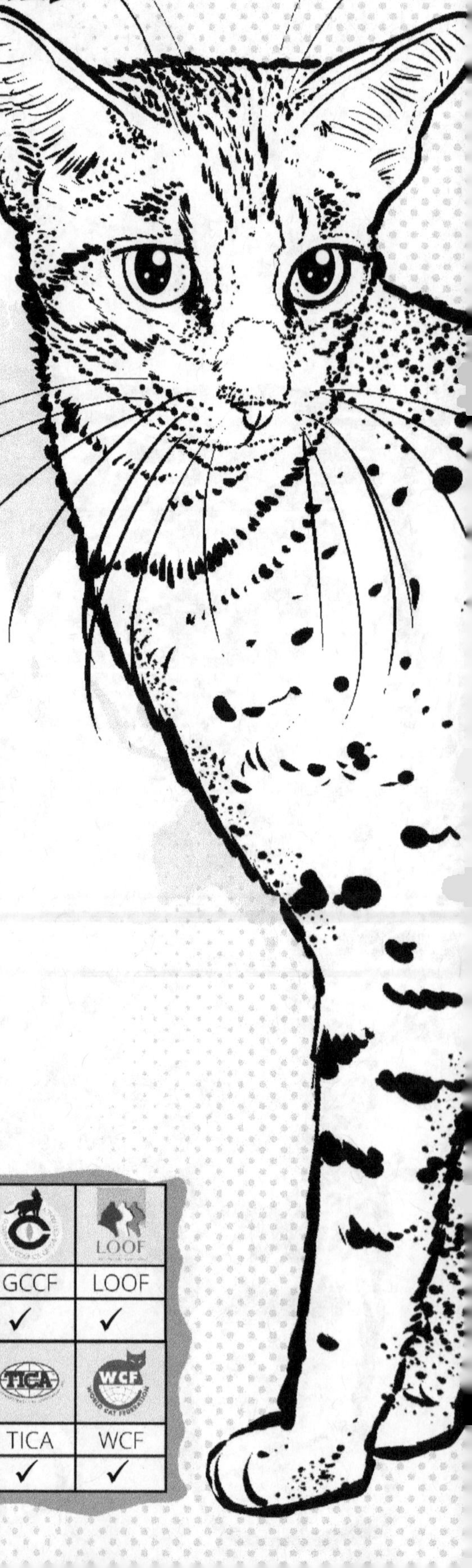

The Ocicat cat was born in 1964 by coincidence in Michigan, USA. In Berkeley, the American breeder Virginia Daly was busy crossing a Siamese cat with an Abyssinian cat. Daly intended to breed a Siamese cat with embroidered Agouti fur - a point cat with the fur of a cougar. The resulting kitten was given the name Tonga. The new cat breed was at this time still called Accicat - derived from the English words, accident, and cat. Tonga was quickly neutered and sold as a pet. Daly, who was enthusiastic about the new cat breed, decided to breed more Accicat cats. Again, she mated Tonga's parents to get another litter of the beautiful cats. Due to her energetic behavior and beautiful appearance, the Offspring was compared to ocelots by the daughter of the breeder, Virginia E. Daly. This is from where the Ocicat gets its name. Soon other breeders were working with enthusiasm on the development of the Accicat cat. Due to an administrative mistake - a simple typo in the breeding standards of the CFA - it was now allowed to cross American Shorthair cats. The result was a very spirited domestic cat with a fur drawing looking like a wild cat.

ACF	ACFA	CAA	CCA-AFC	CCC	CCCA	CFA	FB	FIFe	GCCF	LOOF
✓	✓	✓	✓	✓	✓	✓	✓	✓	✓	✓

NZCF	SACC	TICA	WCF
✓	✓	✓	✓

Selection of the most important cat associations in the world: Australia, Canada, China, France, Germany, New Zealand, South Africa, UK and USA

The Ocicat cat is available in different colors and has a very energetic personality. Like a dog, it obeys its owner and can also be trained. In 1987 the Ocicat cat was recognized by the American CFA as a cat breed.

Recognized since 1966

Breeding start
1980s
United States

DESIGNERCAT

Market price:
800.00 USD–
3.000.00 USD

LIFE

OJOS AZULES SHORTHAIR

The Ojos Azules cat owes its name to its eyes: Ojos Azules is Spanish and translated to mean blue eyes. The unique feature of this shorthaired-cat is the unusual dark blue color of his eyes - which shine even more intensely than those of Siamese cats and thus give the Ojos Azules cat such a real sapphire look. The effect - which produces radiant blue eyes - is due to the reduced or missing melanin production in the iris. This leads to the said optical effect. In the Ojos Azules cat, the mutation is called the Ojos gene. The ancestress of this beautiful cat breed was a cat lady from New Mexico called Cornflower. After some mating of Cornflower with breeding cats, one came to the conclusion that the Ojos gene was dominant. This meant that kittens born purebred, i.e., descendants of parents who both carry the Ojos gene, had numerous hereditary diseases and were born either dead or terminally ill. Since then, Ojos Azules has only mated cats with different eye colors.

Breeding start
1980s
Brazil

Recognized since
1991

LIFE EXPECTATION
10-15
YEARS

ACF	ACFA	CAA	CCA-AFC	CCC	CCCA	CFA	FB	FIFe	GCCF	LOOF

NZCF	SACC	TICA	WCF

Selection of the most important cat associations in the world: Australia, Canada, China, France, Germany, New Zealand, South Africa, UK and USA

ATELIER KAYMAK®
Science
PAPERS

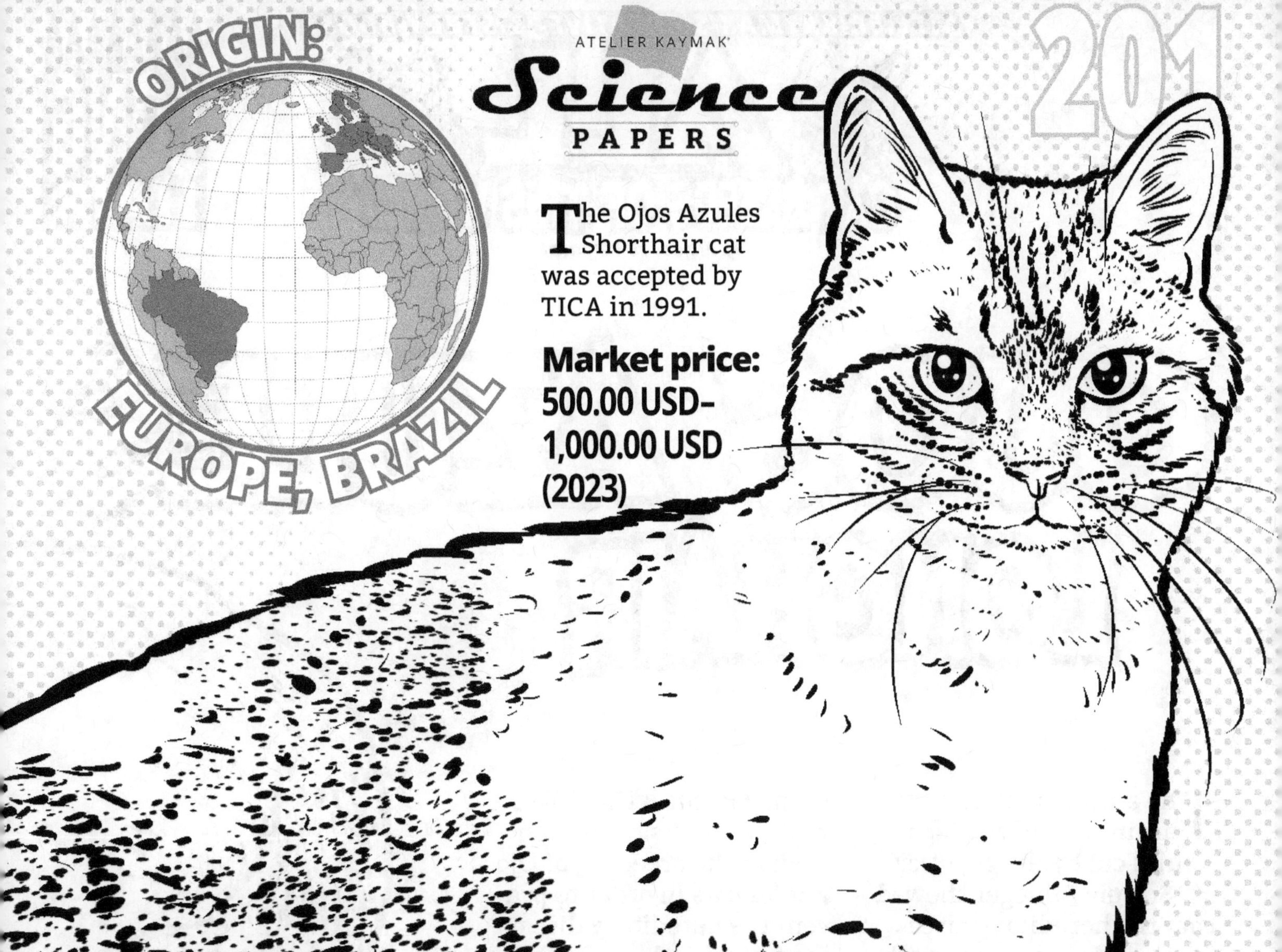

The Ojos Azules Shorthair cat was accepted by TICA in 1991.

Market price: 500.00 USD–1,000.00 USD (2023)

OJOS AZULES LONGHAIR

The analysis of the examination of the Ojos Azules cat by the geneticist Dr. Solveig Pflueger showed that the hereditary diseases usually associated with blue-eyed cats such as deafness and squinting eyes do not occur when parents with different eye colors are paired. This helps to avoid the health problems that usually happen. This led to the fact that since then, parents with different eye colors have been used in the breeding of Ojos Azules cats in order to prevent the hereditary diseases from developing at all. Due to the few breeders in the world, the Ojos Azules cat is still very, very rare to find, and the standards for the cat are therefore still in development. The Ojos Azules longhaired cat was accepted by TICA in 1991.

ACF	ACFA	CAA	CCA-AFC	CCC	CCCA	CFA	FB	FIFe	GCCF	LOOF
										✓

NZCF	SACC	TICA	WCF
		✓	✓

Selection of the most important cat associations in the world: Australia, Canada, China, France, Germany, New Zealand, South Africa, UK and USA

ATELIER KAYMAK
Science
PAPERS

ORIENTAL SHORTHAIR

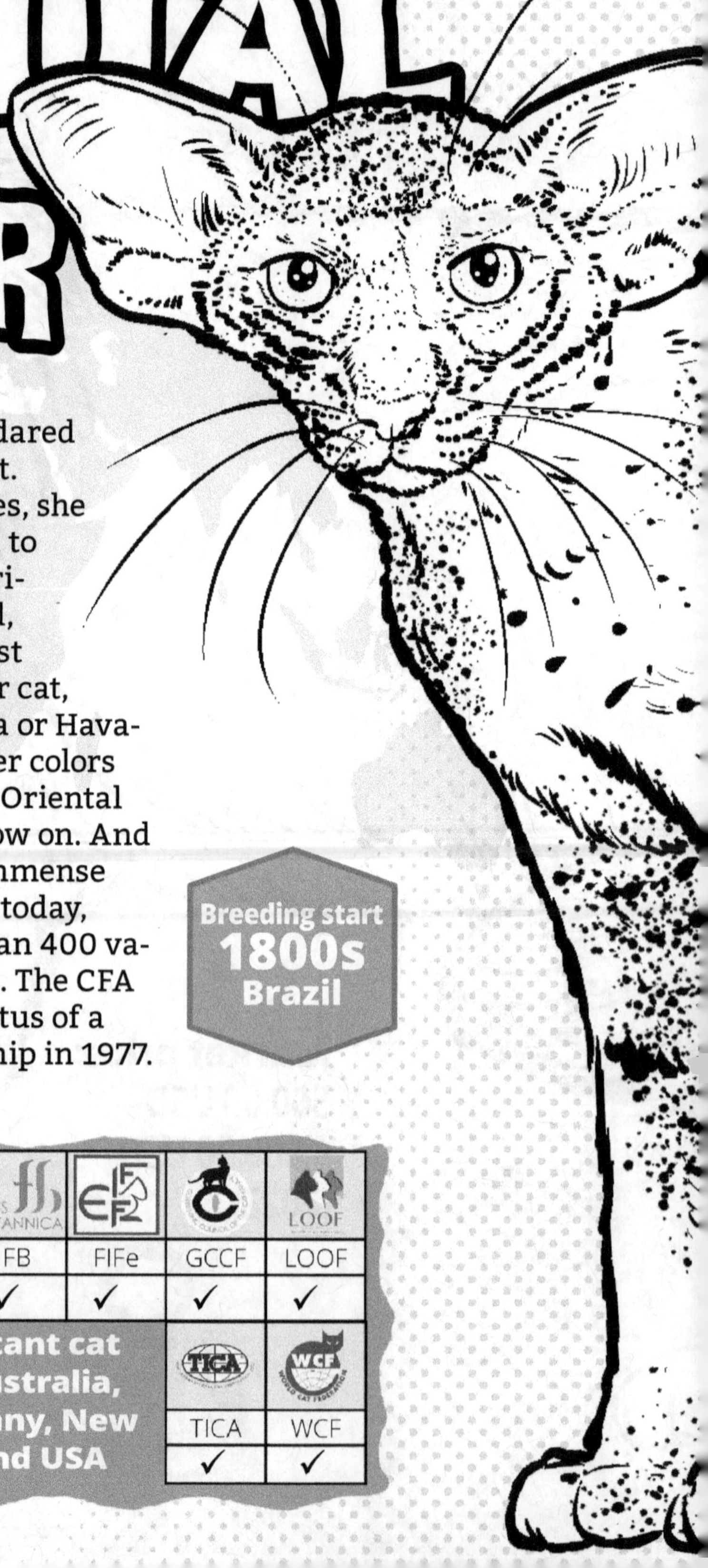

The Oriental Shorthair cat is essentially a Siamese cat, but without the characteristic points - the dark runners at the extremities - which are typical for it. Its breeding history began in the 1950s when many cat breeders increasingly wanted Siamese cats - which should have green instead of blue eyes. They should also now have a short, monochrome brown coat instead of the short, white coat. The demand for further development of traditional color-coding increased. So, in 1951, the English Baroness von Ullman dared to breed such a cat. After some failures, she actually managed to breed a brown variant. This cat breed, considered the first Oriental Shorthair cat, was called Havana or Havana brown. All other colors were classified as Oriental Shorthair from now on. And there is now an immense number of colors: today, there are more than 400 variants in 30 colors. The CFA awarded it the status of a breed championship in 1977.

Breeding start
1800s
Brazil

ACF	ACFA	CAA	CCA-AFC	CCC	CCCA	CFA	FB	FIFe	GCCF	LOOF
✓	✓	✓	✓	✓	✓	✓	✓	✓	✓	✓

NZCF	SACC	TICA	WCF
✓	✓	✓	✓

Selection of the most important cat associations in the world: Australia, Canada, China, France, Germany, New Zealand, South Africa, UK and USA

ORIGIN:

SOUTHEAST ASIA, UK

LIFE EXPECTATION
10-15
YEARS

Market price:
200.00 USD–
3,500.00 USD
(2025)

Recognized since 1977

DESIGNERCAT

Flat-chested Kitten Syndrom (FCKS), Progressive Retinal atrophy (PTR); (with pure white coat) Congenital hearing loss, deafness, disturbances of the

Breeding start
1950s
United States

ORIENTAL LONGHAIR ALSO JAVANESE

The Oriental Longhair cat comes from Great Britain, where breeders tried to replicate the longhaired Turkish Angora cat with longhaired variants of the Oriental Shorthair cats. The result was called British Angora cat or Javanese. This led to some confusion as there was already an American cat breed called Javanese. Many cat associations called British or Turkish long-haired cats simply Angoras. while in continental Europe, the same cat was known as Mandarin, to clarify this. The name was changed to Oriental Longhair in 2002. Like the Oriental Shorthair cat, the Oriental Longhair cat is represented by 281 color variants.

LIFE EXPECTATION
14-16 YEARS

Weight
8-14 lbs

ACF	ACFA	CAA	CCA-AFC	CCC	CCCA	CFA	FB	FIFe	GCCF	LOOF
✓	✓	✓	✓	✓	✓	✓	✓	✓	✓	✓

NZCF	SACC	TICA	WCF
✓	✓	✓	✓

Selection of the most important cat associations in the world: Australia, Canada, China, France, Germany, New Zealand, South Africa, UK and USA

ATELIER KAYMAK
Science
PAPERS

Market price:
400.00 USD–
2,500.00 USD
(2023)

Recognized since 1882

Flat-chested Kitten Syndrom (FCKS)

DESIGNERCAT

PERSIAN

Breeding start
1900
United Kingdom

The Persian cat is one of the oldest cats in the world. Allegedly the Persian cat was brought to Europe in the 17th century by an Italian traveler named Pietro della Valle. The exact origin of the Persian cat is still controversial and probably remains a mystery. According to a legend, King Darius III - who was on the run from the marauding army of Alexander the Great - found refuge on his farm in the mountains where his cats grew their famous long skins to protect themselves from the cold. When the Persian cat was brought to Great Britain, it quickly became one of the most popular cats. It even surpassed the Turkish Angora cat - which had been leading the popularity scale since the early 1900s. At that time, however, the Persian cat looked a little more natural on the face because the typical face that is frequently connected with the race is only a relatively recent development. Persian cats are divided into two categories: the traditional Persian cat - which is based on the classic original and dates back to the 1500s - and the so-called Peke-Faced Persian cat,

ACF	ACFA	CAA	CCA-AFC	CCC	CCCA	CFA	FB	FIFe	GCCF	LOOF
✓	✓	✓	✓	✓	✓	✓	✓	✓	✓	✓

NZCF	SACC	TICA	WCF
✓	✓	✓	✓

Selection of the most important cat associations in the world: Australia, Canada, China, France, Germany, New Zealand, South Africa, UK and USA

ANCIENT ORIGIN

TOP 10

LUXUS

Market price:
US$ 800 to 3,000 (2021)

which only appeared in the late 1950s as a spontaneous mutation and in which the muzzle falls -at, while the nose is small and towers unnaturally high. The CFA moved the peke-faced Persian cat into a separate Category in the years 1958- 1995 because increasing evidence was provided that the peke-face mutation was associated with significant health problems. However, many breeders were used to the flattened face and continued their work. What's more, they concentrated on working out the mutation even more. They produced the so-called Ultra-Typing Persian cat - in which the nose already exceeded the height of the lower eyelids. The cat associations agreed that Persian cats, which exceeded this limit with their nose - are not permitted. The Persian cat is still one of the most popular and best-known breeds in the world.

Recognized since 1889

Market price:
600.00 USD–6,000.00 USD (2025)

Autosomal dominant polycystic kidney disease (ADPKD); (with pure white coat) Congenital hearing loss, deafness, disturbances of the sense of equilibrium, Eye tremor

Breeding start
1994
Russia

PETERBALD

The Peterbald cat belongs to a young cat breed, which only received the international competition status from the AFCA in 2009. It was created in 1994 as part of an experimental breeding program led by the Russian felinologist Olga Miranova in Saint Petersburg. Mrs. Miranova crossed a Donskoy male named Afinogen Myth with the Oriental Shorthair cat and World Champion Radma von Jagerhov. Four elegant, naked cats with extended ears and legs were born as a result of two matings. The four kittens - Mandarin iz Murino, Mus cat iz Murino, Nezhenka iz Murino, and Nocturne iz Murino - had some downy hair but were otherwise bald and remained hairless afterward. They formed the basis of the Peterbald breeding. The demanding Peterbald cat is still quite rare to find and therefore costs a fortune. It is also not known which crossings are necessary to produce hairless Peterbald kittens. Therefore, experiments are still being conducted to determine the exact gene responsible for it. TICA accepted the Peterbald cat in 1997, the WCF in 2003. In countries like Germany, the animal protection law according to § 11b (Qualzucht-Paragraph) prohibits the breeding of the Peterbald cat.

ACF	ACFA	CAA	CCA-AFC	CCC	CCCA	CFA	FB	FIFe	GCCF	LOOF
✓	✓	✓	✓				✓	✓		✓

NZCF	SACC	TICA	WCF
	✓	✓	✓

Selection of the most important cat associations in the world: Australia, Canada, China, France, Germany, New Zealand, South Africa, UK and USA

ATELIER KAYMAK
211
ORIGIN:
ASIA, RUSSIA
Recognized since 2009
Market price:
1,800.00 USD–
5,500.00 USD
(2025)
LIFE EXPECTATION
10-15
YEARS
LUXUS
HAIRLESS
Progressive Retinal

PIXIEBOB SHORTHAIR

The Pixiebob cat, which looks like a mini lynx, is characterized by her size, her tail, and her toes. The peculiar breed was developed in 1986 in the Pacific Northwest of the United States by Carol Ann Brewer after she adopted an American stumpy tail male named Keba, which was so large that it reached her to the knees. Keba mated with her neighbor's cat and produced an offspring that stood out from the other kittens in litter size alone: the unique-looking American stump tail and polydactyly cat named Pixie - enjoyed life! Knowing the uniqueness of Pixie, Brewers and other breeders specializing in wild-looking cats set out to continue the development of this new breed. The American TICA finally recognized the cat breed in 1998. According to the breed standards of the TICA, the Pixiebob cat is allowed a maximum of 7 toes per paw in competi tions. This is more than enough: 28 toes are eight toes more than normal cats have!

BOBTAIL CAT

Breeding start
1986
United States

ACF	ACFA	CAA	CCA-AFC	CCC	CCCA	CFA	FB	FIFe	GCCF	LOOF
✓	✓	✓	✓				✓	✓		✓

NZCF	SACC	TICA	WCF
	✓	✓	✓

Selection of the most important cat associations in the world: Australia, Canada, China, France, Germany, New Zealand, South Africa, UK and USA

ATELIER KAYMAK®

Science PAPERS

LIFE EXPECTATION
4-10 YEARS

Recognized since 1998

Market price:
200.00 USD–
5,500.00 USD
(2025)

BIG CAT

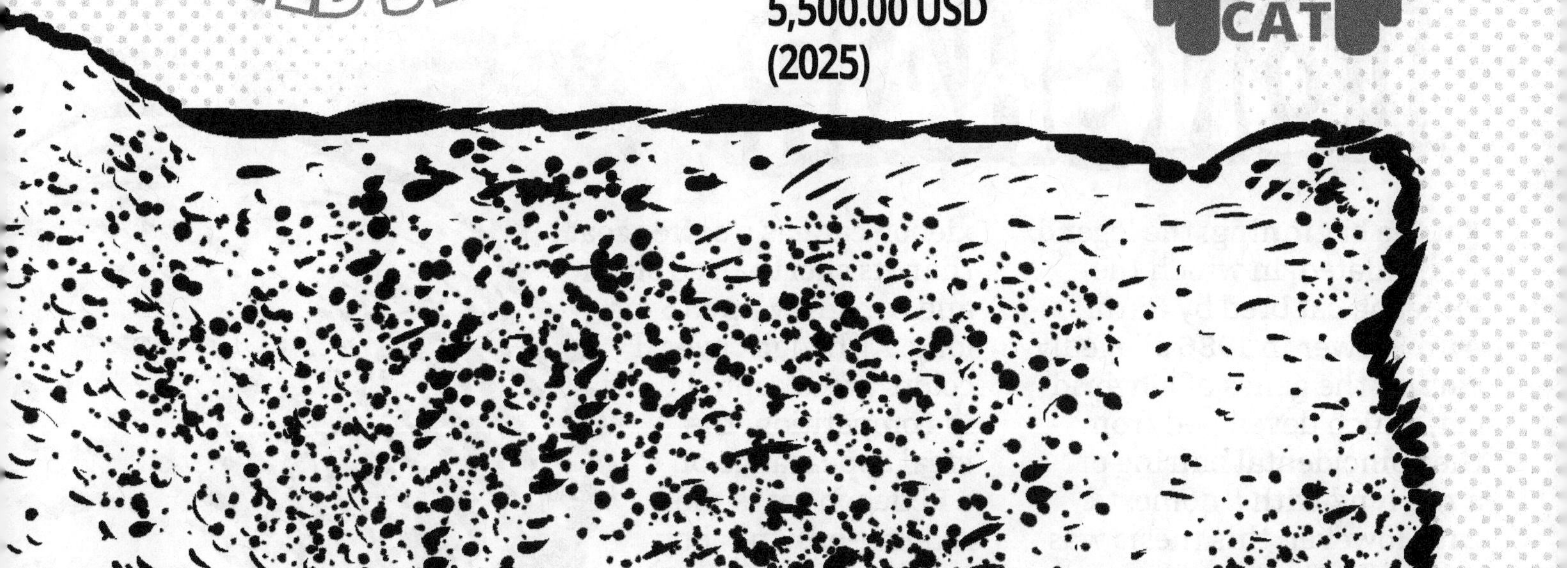

Weight

PIXIEBOB LONGHAIR

Breeding start
1994
Russia

BIG CAT

In the beginning, the legend circulated, in which the Pixiebob cat bred by Carol Ann Brewer in 1986 allegedly carries the genes of a hybrid cat, which developed from the coincidental pairing of a red lynx with a domestic cat. However, this thesis was scientifically refuted in the mid-1990s by DNA comparison of Pixiebob cats and Red Lynxes. The Pixiebob cat is, therefore, genetically a pure domestic cat. But it is proven that Brewer is said to have raved about a supernatural appearance, a pixie when she saw the lynx-like kitten. The long-haired version of the Pixiebob cat has a softer coat than its shorthaired sister and can grow up to 5 cm long. Although several colors are accepted in competition, the ideal appearance of a Pixiebob cat is the reddish-brown tabby pattern. Thus the cat resembles in appearance the red lynx living near the coast, which was the godfather for the cat breed. Pixiebob cats have a sturdy, muscular body and a pronounced rolling gait - similar to that of wild cats. Some larger specimens weigh up to 5 kg.

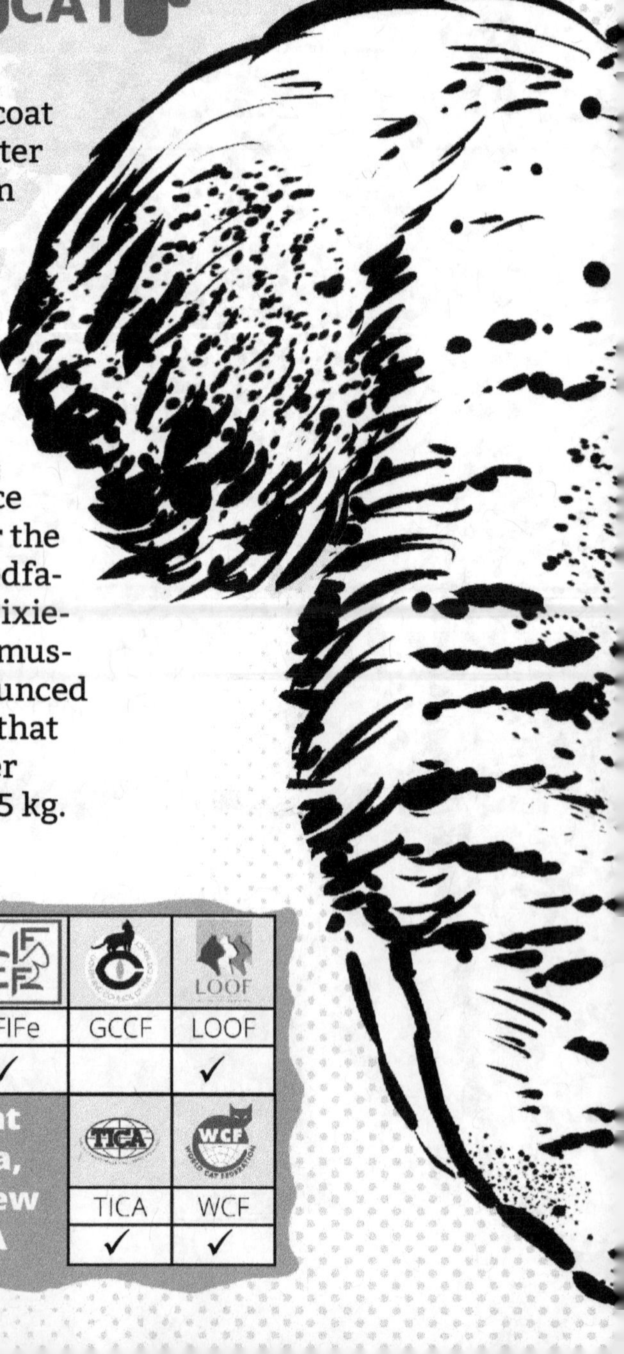

ACF	ACFA	CAA	CCA-AFC	CCC	CCCA	CFA	FB	FIFe	GCCF	LOOF
✓	✓	✓	✓				✓	✓		✓

NZCF	SACC	TICA	WCF
	✓	✓	✓

Selection of the most important cat associations in the world: Australia, Canada, China, France, Germany, New Zealand, South Africa, UK and USA

Pixiebob cats mature slowly and take almost four years to reach adulthood. The tail length can vary from very short to long. The tufts of fur on ears and hair grow down along the cheeks and look a little like sideburns - reminiscent of the cartoon figure WOLVERINE and enhancing the wild aura.

Recognized since 2009

Market price:
200.00 USD–
5,500.00 USD
(2025)

BOBTAIL CAT

LIFE EXPECTATION

POLYCOON

BIG CAT

Normal cats have five toes on the front paws, while the hind paws have four toes. A polydactyl mutation - the multi-fingers - consists, as the name suggests, essentially of additional toes on the paws - but this has no significant effect on the cat's life. The mutation was first registered in the eastern United States, with Maine Coon cats accounting for a particularly high proportion of about 40 percent. Since the mutation was considered a defect, selective breeding was used to avoid additional toe formation. In recent years, however, breeders have come to appreciate the mutation. The Polycoon cat became famous through the famous writer Ernest Hemingway when he lived in Florida on the island Key West in the 1930s. He was given a polydactyl cat by a captain. Snowball, as Hemingway's tomcat was called, quickly mated with the native cats. Hemingway loved these cats so much that they named themselves after his name. The author had fun and named every single newborn of his cats after a famous actor. Today about 60 Hemingway cats still live on the island. The well-protected cats are considered the heir of the famous author and are therefore not sold. In countries like Germany, the animal protection law prohibits the breeding of the Polycoon cat.

Breeding start 1950s United States

LIFE EXPECTATION 14-16 YEARS

Weight 8-18 lbs

ACF	ACFA	CAA	CCA-AFC	CCC	CCCA	CFA	FB	FIFe	GCCF	LOOF

NZCF	SACC	TICA	WCF
		✓	

Selection of the most important cat associations in the world: Australia, Canada, China, France, Germany, New Zealand, South Africa, UK and USA

Hypertrophic cardiomyopathy (HCM), Spinale Muskelatrophien (SMA); (with pure white coat) Congenital hearing loss, deafness, disturbances of the sense of equilibrium, Eye tremor

ATELIER KAYMAK
217
ORIGIN:
EUROPE, UNITED STATES
Recognized since 1882
Market price:
100.00 USD–
1,300.00 USD
(2023)

RAGAMUFFIN

As an Offspring of the Ragdoll cat, the Ragamuffin cat is the product of a number of breeders who left The International Ragdoll cat Association IRCA, founded by Carol Ann Brewer, because of various disagreements. The Ragamuffin cat, also known as the modern Ragdoll cat, was developed in 1994 from crossing IRCA-Ragdolls with Oriental Longhair and Persian cats. The newborn cats were so sweet and tender that they got the nickname cherubim (supernatural being, also angel). Like the Ragdoll cat, the Ragamuffin cat is exceptionally tame and has no problem being taken on the arm and stroked. In CFA- cat shows all color variations of the big, fluffy, and uncomplicated cat are allowed except Colorpoint cats.

Breeding start
1960s
United States

ACF	ACFA	CAA	CCA-AFC	CCC	CCCA	CFA	FB	FIFe	GCCF	LOOF
	✓	✓	✓			✓	✓	✓		

NZCF	SACC	TICA	WCF
			✓

Selection of the most important cat associations in the world: Australia, Canada, China, France, Germany, New Zealand, South Africa, UK and USA

ATELIER KAYMAK
Science
PAPERS

Recognized since 2003

Market price:
200.00 USD–3,000.00 USD (2025)

LIFE EXPECTATION
12-18
YEARS

SHORT HAIR

Breeding start
1960s
United States

RAGDOLL

BIG CAT

The Ragdoll cat is a cross between a Burmese cat and a Persian cat which was injured in a traffic accident in the 1960s. The resulting litter included beautiful looking kittens with a calm temperament. Legend has it that the newborns were said not to be able to feel pain, which led to their name: Ragdoll. In fact, the Ragdoll cat looks like a lifeless doll when picked up. Of course, this was a great breeding success, because everyone was immediately taken with the apparently insensitive cat. Quickly the suitable name was also found - Ragdoll - which became for each cat fan the term for the new cat race. However, the popularity of the cat moved the then breeder Carol Ann Brewer to a step that made her unpopular among breeders: contrary to the principles of the associations, she had the name Ragdoll registered as a trademark at the patent office in the early 1970s. As a result, the use of the name was protected under trademark law from then on. Subsequently, every breeder of Ragdoll cats would have had to pay license fees to Brewer. But the enterprising Brewer went even further: she began to advertise the Ragdoll cat as a toy cat, which made the cat associations even more upset. Brewer wanted to spread the legend of the pain-free cat. As proof she even threw one of her Ragdoll cats across the room during a television interview. The animal did not react very impressed and seemed to support Brewers bold claim.

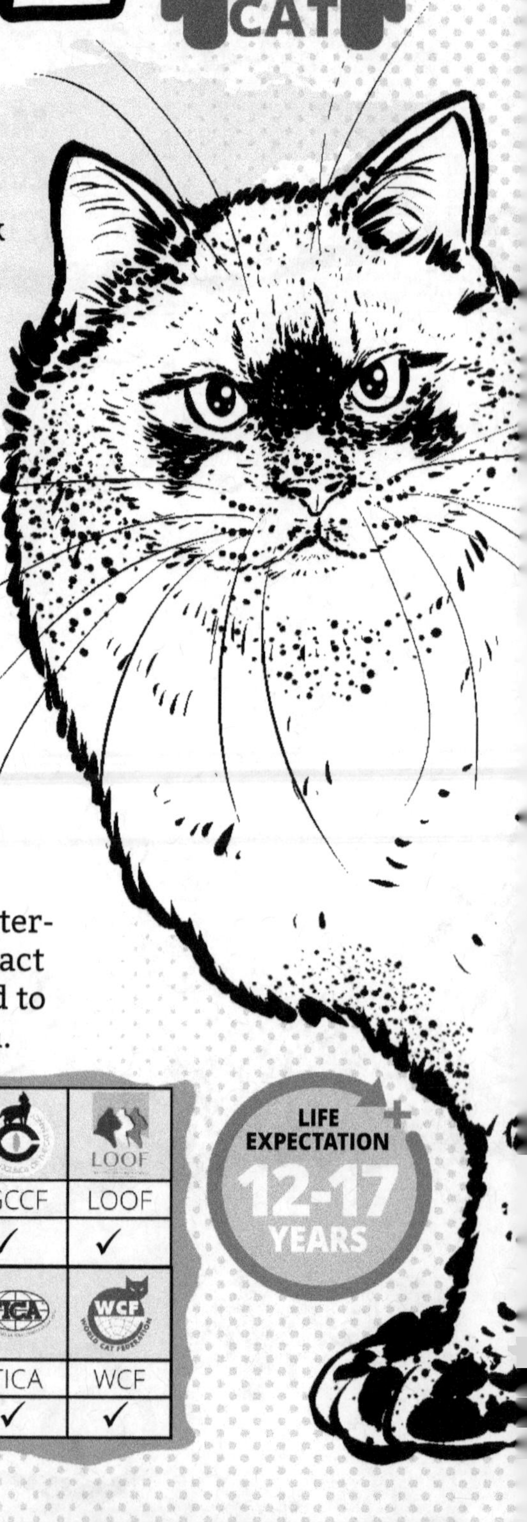

ACF	ACFA	CAA	CCA-AFC	CCC	CCCA	CFA	FB	FIFe	GCCF	LOOF
✓	✓	✓	✓	✓	✓	✓	✓	✓	✓	✓

NZCF	SACC	TICA	WCF
✓	✓	✓	✓

Selection of the most important cat associations in the world: Australia, Canada, China, France, Germany, New Zealand, South Africa, UK and USA

LIFE EXPECTATION
12-17
YEARS

More likely, however, was that Mrs. Brewer had created even more tame offspring by crossing an already very gentle Burmese cat with an equally tame Persian. Brewer had now created enemies for himself. As a result she was boycotted by all cat associations and neither the breeder nor her Ragdoll cats were admitted to shows. Brewer did not let himself be put off and simply founded her own cat register: The International Ragdoll cat Association (IRCA). But for the associations neither Brewer nor the Ragdoll cat existed. Brewer did their own thing and nobody was interested in each other. It was only in 2005, three decades later, that the trademark protection of the Ragdoll cat expired. Since then the Ragdoll cat is recognized as own cat race by all federations.

Market price:
200.00 USD-
1,500.00 USD
(2025)

Recognized since 2005

TOP 10

LUXUS

Hypertrophic cardiomyopathy (HCM), Urinary stones; (with pure white coat) Congenital hearing

RUSSIAN BLUE

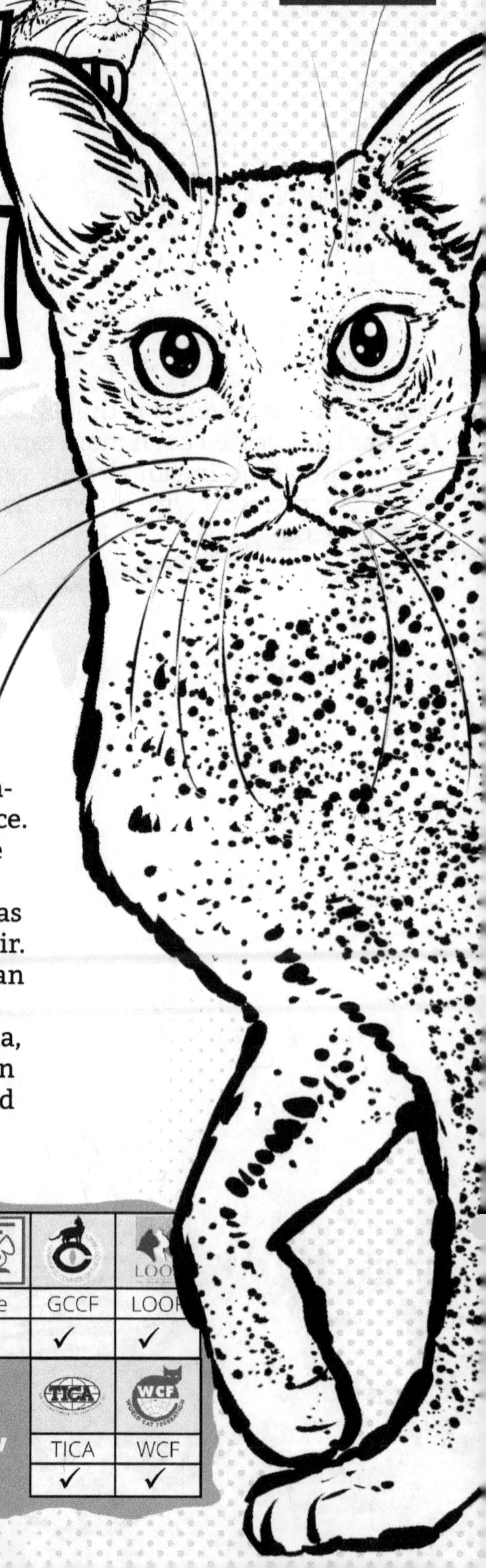

Initially, the cat associations evaluated all blue-haired cats together in one Category - until the Russian blue cat got its own class: In 1912, the associations introduced the Foreign blues Category, which only applied to the Russian blue cat. In the first half of the 20th century, the popularity of this cat increased rapidly - its breeding was on a large scale: simultaneously, it was further developed in Great Britain, Scandinavia, and the USA. American breeders combined the results of their work with the two other lines and created the modern Russian blue cat as we know it today: with distinctive dark emerald green eyes, a uniquely shiny and soft blue-grey coat, and a strange-looking smile on her face. The dense, plush fur of the Russian blue cat is created by the undercoat, which has the same length of coat hair. Since the 1960s, the Russian blue cat has been accepted by associations in Australia, South Africa, and the UK in the colors arctic, white, and black in their own classes.

ACF	ACFA	CAA	CCA-AFC	CCC	CCCA	CFA	FB	FIFe	GCCF	LOOF
✓	✓	✓	✓	✓	✓	✓	✓	✓	✓	✓

NZCF	SACC	TICA	WCF
✓	✓	✓	✓

Selection of the most important cat associations in the world: Australia, Canada, China, France, Germany, New Zealand, South Africa, UK and USA

ATELIER KAYMAK®
Science
PAPERS

ORIGIN:

ASIA, UNITED KINGDOM

LUXUS

Recognized since 1937

Market price:
500.00 USD–
2,500.00 USD
(2025)

ANCIENT ORIGIN

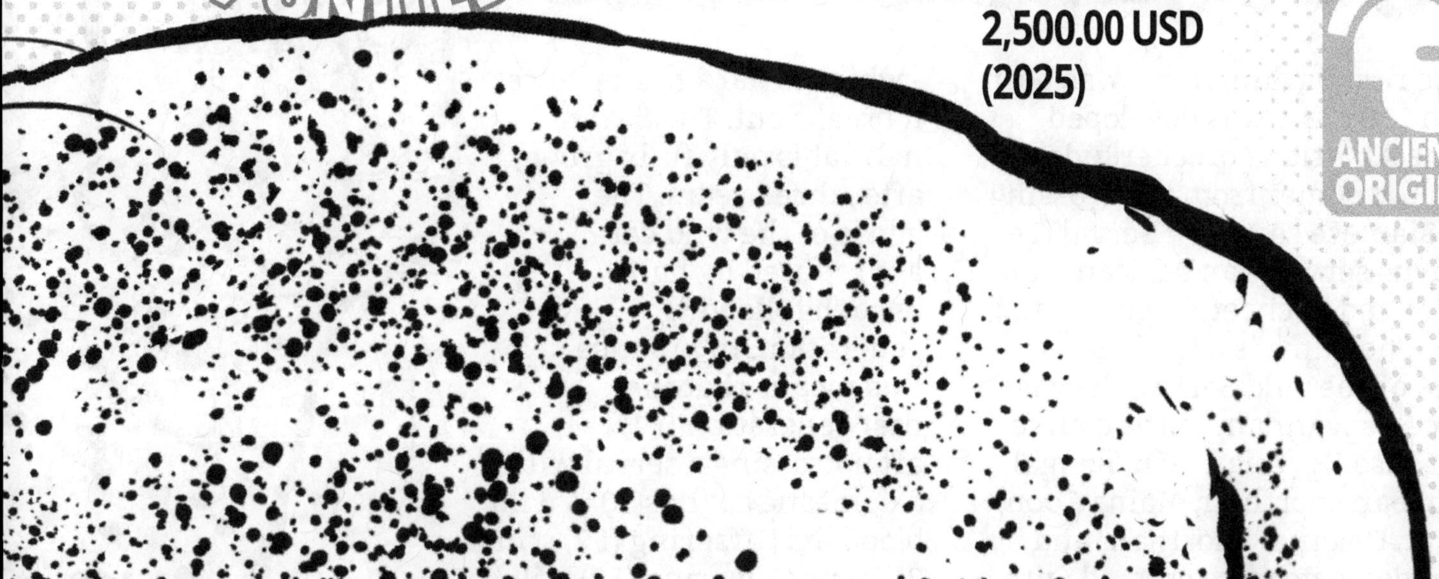

Breeding start
1800s
United Kingdom

LIFE

SHORT HAIR

Breeding start
1986
United States

SAVANNAH

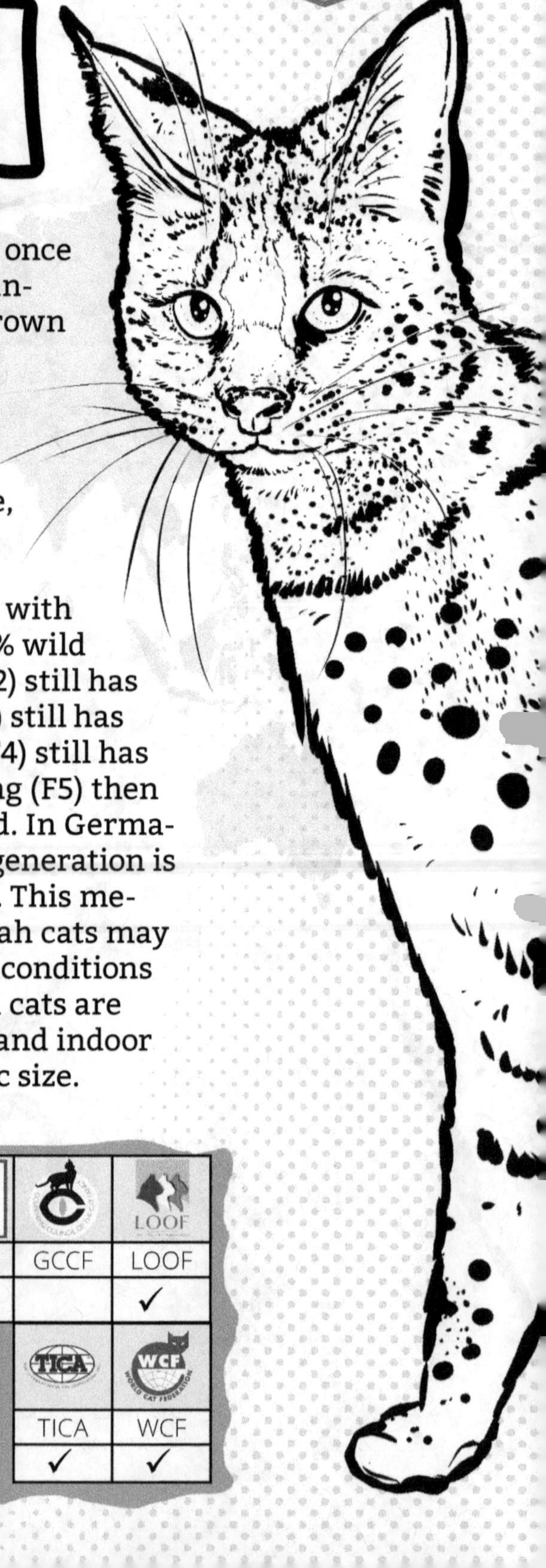

The first Savannah cat was born in 1986 and was developed by the American breeder Judee Frank. It comes from the crossing of a Siamese cat and a Serval (Leptailurus serval) - an African wild cat related to cheetah, puma, and lynx. The name is a homage to the home of the wild Serval, the vast African savannah. In the course of time, also Egyptian Mau, Bengal, European shorthair, Maine Coon, Ocicat, Oriental Shorthair, and Serengeti cats were crossed with the Savannah cat. The wild genes of the Serval are noticeable: the large Savannah cat has long, powerful legs - with which it can jump over 2 meters from a standing position. She has a small head, big ears, and the usual spotting pattern on the coat for wild cats. The friendly, sociable, and playful cat has a strong hunting instinct - which is hard to stop once it breaks out. The Savannah cat is only fully grown after three years. The genes of the wild cat are reduced by half in each following generation. In practice, this means that the first generation (F1) after crossing a serval with a domestic cat has 50 % wild blood; its Offspring (F2) still has 25 %; its Offspring (F3) still has 12.5 %; its Offspring (F4) still has 6.25 % and its Offspring (F5) then only has 3 % wild blood. In Germany, the first to fourth generation is subject to registration. This means that these Savannah cats may only be kept if certain conditions for the keeping of wild cats are fulfilled: e.g., outdoor and indoor enclosures of a specific size.

ACF	ACFA	CAA	CCA-AFC	CCC	CCCA	CFA	FB	FIFe	GCCF	LOOF
							✓	✓		✓

NZCF	SACC	TICA	WCF
		✓	✓

Selection of the most important cat associations in the world: Australia, Canada, China, France, Germany, New Zealand, South Africa, UK and USA

From the 5th generation onwards, the cat is considered a typical domestic cat (in Switzerland, this applies from the 3rd generation onward). The Savannah cat is a graceful cat that is sought after everywhere. And that makes it one of the most expensive cats in the world: you have to spend up to 50,000 US$ if you want the luxury cat. TICA recognized the breed in 2001 and awarded it championship status in 2012.

Recognized since 2001

BIG CAT

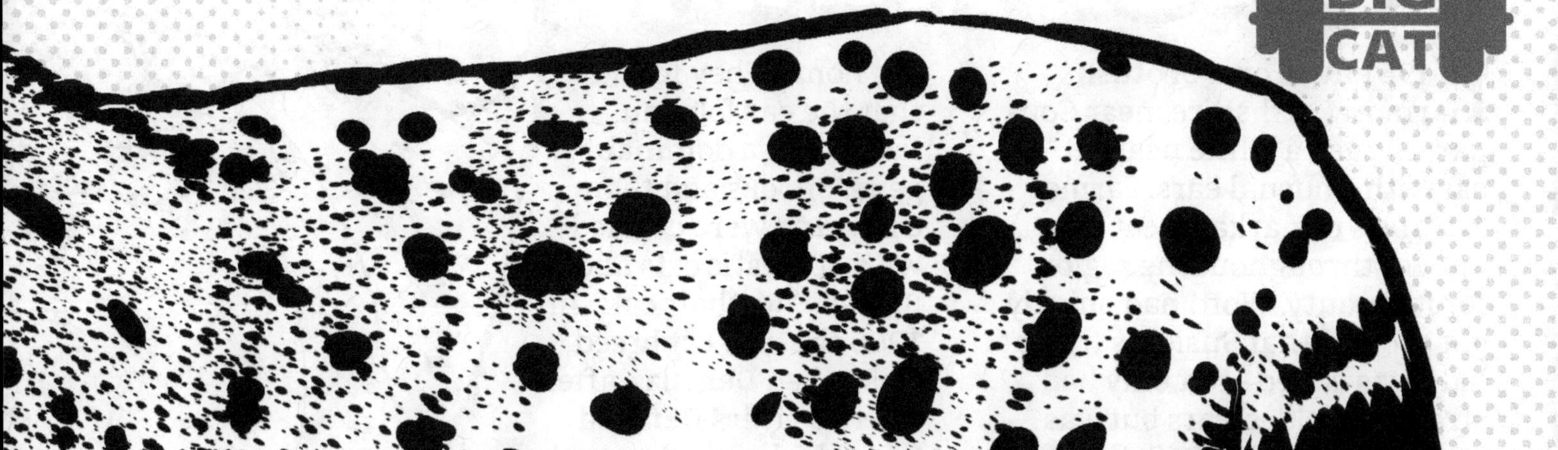

HYBRID

Market price:
200.00 USD–
50,000.00 USD
(2025)

LIFE

SCOTTISH FOLD SHORTHAIR

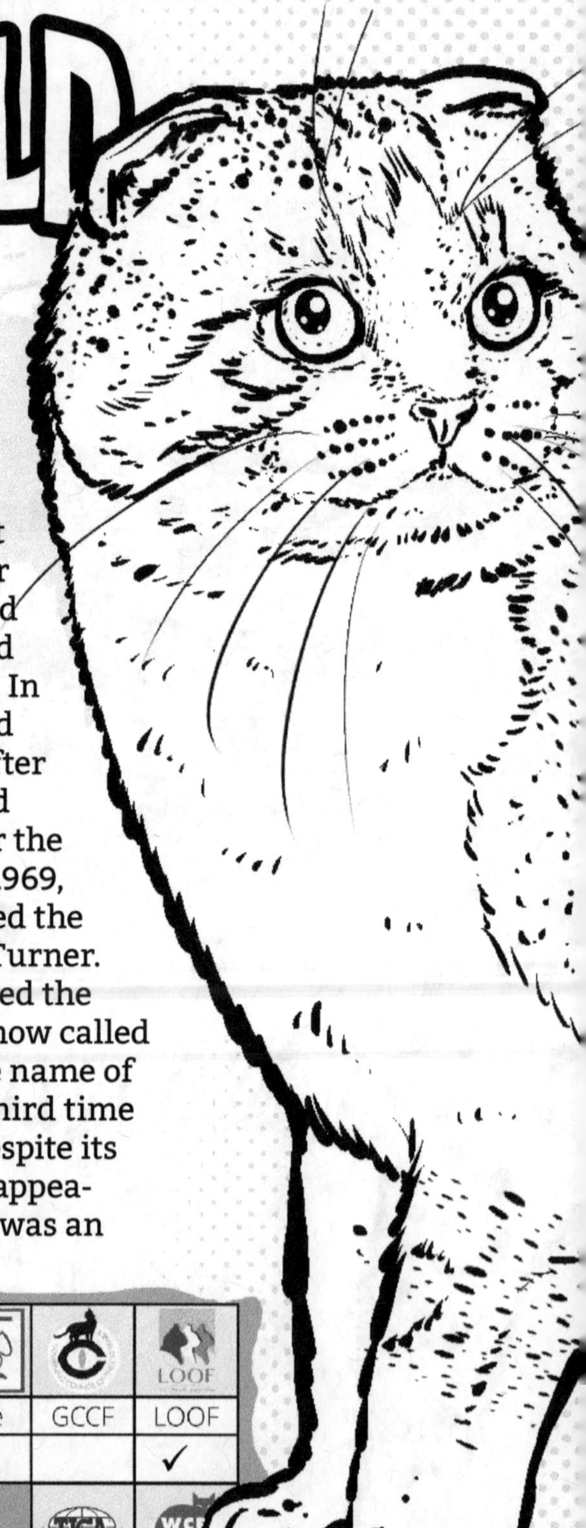

In 1961 lived on a Scottish farm in Perthshire, near Coupar Angus - a white peasant cat with unusual ears. Farmer McCrae's remarkable cat was known throughout the region for its beauty. Word had quickly got around that Susie, as the cat lady was called - not only had strangely folded ears but was also pregnant. In 1963, this also reached William and Mary Ross, passionate cat breeders. They had become curious about Susie and paid Mr. McCrae a visit. They offered the farmer some money for one of the puppies, which of course, the farmer did not refuse. After choosing a female, they christened her Snooks (from the billiards game snooker) and drove her home. They found that the ear fold mutation was a dominant gene. Snooks and their Offspring were crossed with domestic cats and British Shorthair cats. In 1966 the new cat breed was called Denisla - after the two rivers Den and Isla, which -owed near the Mr. and Mrs. Ross. In 1969, the Ross couple enlisted the help of geneticist Pat Turner. Together they developed the cat breed, which was now called Lops, and changed the name of the cat breed for the third time to Scottish fold cat. Despite its beautiful and unique appearance, however, there was an

ACF	ACFA	CAA	CCA-AFC	CCC	CCCA	CFA	FB	FIFe	GCCF	LOOF
✓	✓	✓	✓	✓	✓	✓				✓

NZCF	SACC	TICA	WCF
✓	✓	✓	✓

Selection of the most important cat associations in the world: Australia, Canada, China, France, Germany, New Zealand, South Africa, UK and USA

agreement in England that folding ears were a malformation that caused health problems. As a result, the GCCF revoked the recognition as a new breed that it had issued in 1966. Once again, there were liberal breeders in the USA who guaranteed the continuation of stopped breeding and ensured the continuation of the Scottish fold cat. Sally Wolfe founded the International Scottish Folds Association in 1978 and further developed the breed with Scottish and American Shorthair cats. In countries like Germany, the animal protection law prohibits the breeding of the Scottish fold cat.

Breeding start
1960s
United Kingdom

LUXUS

LIFE EXPECTATION
10-15
YEARS

Recognized since
1966

Market price:
200.00 USD–4,500.00 USD (2025)

Autosomal dominant polycystic kidney disease (ADPKD), Hypertrophic cardiomyopathy (HCM), Osteochondrodysplasia (OCD); (with pure white coat) Congenital hearing loss, deafness, disturbances of the sense of equilibrium, Eye tremor

SCOTTISH FOLD LONGHAIR

The long-haired Scottish fold cat is better known as Highland-Fold or Coupari in places. The history of the breeding cat begins in 1961, when a cat breeder from Perthshire in Scotland became aware of the white cat of his neighbor. The farmer's cat Susie, had strangely curled ears which gave her an owlish appearance. Susie was therefore a local celebrity and everyone wanted to see her. So William Ross also became aware of the farmer cat Susie. He urged the farmer to sell him the pregnant cat which he refused. But he promised to give him some of Susie's kittens. Ross registered both kittens at the GCCF and started breeding. For a long time the existence of the long-haired Scottish fold cat was overshadowed by its short-haired counterpart. The long fur concealed the cat's folded ears and made them look earless. This changed in 1980 when Hazel Swadberg made the long-haired Scottish fold cat popular at cat shows in the USA. In 1986 the TICA recognized the long-haired Scottish fold cat as an independent breed.

ACF	ACFA	CAA	CCA-AFC	CCC	CCCA	CFA	FB	FIFe	GCCF	LOOF
✓			✓	✓	✓	✓				✓

NZCF	SACC	TICA	WCF
✓	✓	✓	✓

Selection of the most important cat associations in the world: Australia, Canada, China, France, Germany, New Zealand, South Africa, UK and USA

In countries like Germany the animal protection law prohibits the breeding of the Scottish fold cat.

LUXUS

Recognized since 1986

Breeding start
1950s
United Kingdom

Market price:
200.00 USD–
4,500.00 USD
(2025)

LIFE EXPECTATION
13-15
YEARS

Autosomal dominant polycystic kidney disease (ADPKD), Hypertrophic cardiomyopathy (HCM), Osteochondrodysplasia (OCD); (with pure white coat) Congenital hearing loss, deafness, disturbances of the sense of equilibrium, Eye tremor

SCOTTISH SHORTHAIR

Breeding start **1950s** United Kingdom

LIFE EXPECTATION **10-15** YEARS

The Scottish Shorthair cat is identical to the Scottish Straight cat. We list it here for the sake of completeness because it is listed by some federations under this name. Scottish Shorthair cats are essentially Scottish fold cats whose ears have grown normally (50% of the litter). These special cats with the typical ear fold mutation can have either one or two copies of the fold gene Fd - FdFd or Fdfd, while cats with smooth ears have two copies of the normal smooth ears gene - fdfd. Since breeding homozygous cats with the wrinkle gene FdFd is associated with the breeding of cats with homozygous (clean ears), there is a considerable risk of health problems. The only generally accepted breeding results in a 50% chance of producing heterozygous (mixed) folding ear cats. This means that for every Scottish fold cat, there is a counterpart with straight ears. Perhaps, given the inevitable abundance of cats, it is only natural to give them competitive status. In Australia, the Scottish Shorthair cat is clearly classified: it received competition status from the ACF in 2002.

ACF	ACFA	CAA	CCA-AFC	CCC	CCCA	CFA	FB	FIFe	GCCF	LOOF
✓			✓	✓		✓				✓

NZCF	SACC	TICA	WCF
✓	✓	✓	✓

Selection of the most important cat associations in the world: Australia, Canada, China, France, Germany, New Zealand, South Africa, UK and USA

ORIGIN:

UNITED KINGDOM

Recognized since 1966

Market price:
800.00 USD–2,000.00 USD (2023)

Autosomal dominant polycystic kidney disease (AD-PKD), Hypertrophic cardiomyopathy (HCM), Osteochondrodysplasia (OCD); (with pure white

SCOTTISH LONGHAIR

Breeding start **1960s** United Kingdom

LIFE EXPECTATION **10-15** YEARS

The Scottish Longhair cat has a particularly innocent-looking and open expression. Owners of Scottish Longhair cats will sooner or later discover some special features in their cats, such as the tendency to be overweight. Or their strange eating behavior because Scottish cats use their paws to eat and drink. They love to sit with their legs stretched out on their bottom and to attentively and calmly register what is happening in their environment. While her front paws are lying on her stomach, she looks, depending on the situation, either interested or bored at the goings-on of her owner, who, after long, silent observation phases of his cat, feels a little insecure, always measured and evaluated by his own cat. Scottish fold cats show another scurrility: they sleep on their back! Then they lie there with their arms and legs stretched out and their head tilted backward, just like a discarded doll. Their intelligence is legendary: they learn very quickly to crack open cupboards and doors to get at treats. Or they follow their owner tirelessly through the whole house and create a spooky Pet Cemetery atmosphere (you know, the cat horror novel by Stephen King). And Scottish longhaired cats are very chatty because when they are on the move, they babble like a waterfall. Often, they tell their owner (who should listen patiently) in soft and quiet tones about all the exciting events of the day. The longhaired version of the Scottish Shorthair cat was accepted by the Australian cat Association ACF in 2002.

ACF	ACFA	CAA	CCA-AFC	CCC	CCCA	CFA	FB	FIFe	GCCF	LOOF
✓	✓	✓		✓	✓	✓	✓	✓		✓

NZCF	SACC	TICA	WCF
✓	✓	✓	✓

Selection of the most important cat associations in the world: Australia, Canada, China, France, Germany, New Zealand, South Africa, UK and USA

ATELIER KAYMAK®

Science

PAPERS

233

ORIGIN:

UNITED KINGDOM

Recognized since 2002

Market price:
500.00 USD–800.00 USD (2023)

Autosomal dominant polycystic kidney disease (ADPKD), Hypertrophic cardiomyopathy (HCM), Osteochondrodysplasia (OCD); (with pure white coat) Congenital hearing loss, deaf

SCOTTISH STRAIGHT

Breeding start
1960s
United Kingdom

The Scottish Straight cat is identical to the Scottish Shorthair cat. We list them here for the sake of completeness because some federations use this name. Scottish Straight cats are essentially Scottish fold cats whose ears have grown normally (50% of the litter). Because the ear-folding gene is dominant and crossing two Scottish fold cats causes health problems, these special cats are usually crossed with American or English Shorthair cats. This means that half of the breeding cats continue to develop their distinctive folding ears, while the other half have normal, straight ears. All live Scottish fold cats - including Straight Scottish cats - can be traced back to the Dame Susie from farmer McCrae. In temperament, Scottish Straight cats are amiable and curious. They have big round eyes, are pleasantly quiet, and never too pushy. Some compare their clever facial expression (and thinking apparatus) to that of an owl. It is well known that Straight Scottish cats like to relax on their hind legs and stand upright.

ACF	ACFA	CAA	CCA-AFC	CCC	CCCA	CFA	FB	FIFe	GCCF	LOOF
			✓							

NZCF	SACC	TICA	WCF
			✓

Selection of the most important cat associations in the world: Australia, Canada, China, France, Germany, New Zealand, South Africa, UK and USA

ATELIER KAYMAK
Science
PAPERS
235
ORIGIN:
UNITED KINGDOM
Recognized since 2002
LIFE EXPECTATION
12-15
YEARS

Breeding start
1980s
United States

SELKIRK REX SHORTHAIR

The first Selkirk Rex cat was born in 1987 in Sheridan, Montana, USA. At For Pet's Sake shelter, a cat with three legs had kittens. She had previously been trapped and was very injured when she was found. The recovered mother gave birth to healthy kittens among which was a female newborn with curly hair. The unusual kitten landed at the well-known Persian cat breeder Jeri Newman via two previous owners (in the USA, everything goes very fast if something looks only approximately unusual). Mrs. Newman inspected the kitten, aptly named her Miss Depesto of Noface, and kept her. A few months later, she mated it to her champion Black Persian, a Persian male. In the litter two months later, six tiny kittens saw the light of day. It was striking that one half of the Offspring of Miss Depesto of Noface, as expected, had little curls in their fur, but the other half had a smooth fur coat. Short insertion: The curls in the coat are caused by a mutation in the KRT71 gene, which encodes the protein keratin 71 and thus controls the development of the

ACF	ACFA	CAA	CCA-AFC	CCC	CCCA	CFA	FB	FIFe	GCCF	LOOF
✓	✓	✓	✓	✓	✓	✓	✓	✓	✓	✓

NZCF	SACC	TICA	WC
✓	✓	✓	✓

Selection of the most important cat associations in the world: Australia, Canada, China, France, Germany, New Zealand, South Africa, UK and USA

LIFE EXPECTATION
10-15
YEARS

hair. A malfunction can cause permanent waves or curls. In the case of a complete loss, e.g., the Sphynx cat, the hair breaks off and falls out. Although all Rex cats carry a mutated KRT71 gene - the Selkirk Rex cat has another mutation of the gene - which distinguishes it from the other Rex cats. End of insertion. In any case, it was now clear that autosomal dominant inheritance made it possible to obtain a number of different fur types through different breeding methods: homozygous producers (both parents with the mutated gene) produced smaller and denser curls while heterozygous producers (only one parent with the mutated gene) produced stronger and wavier curls.

Market price:
150.00 USD–
2,800.00 USD
(2023)

Recognized since
1992

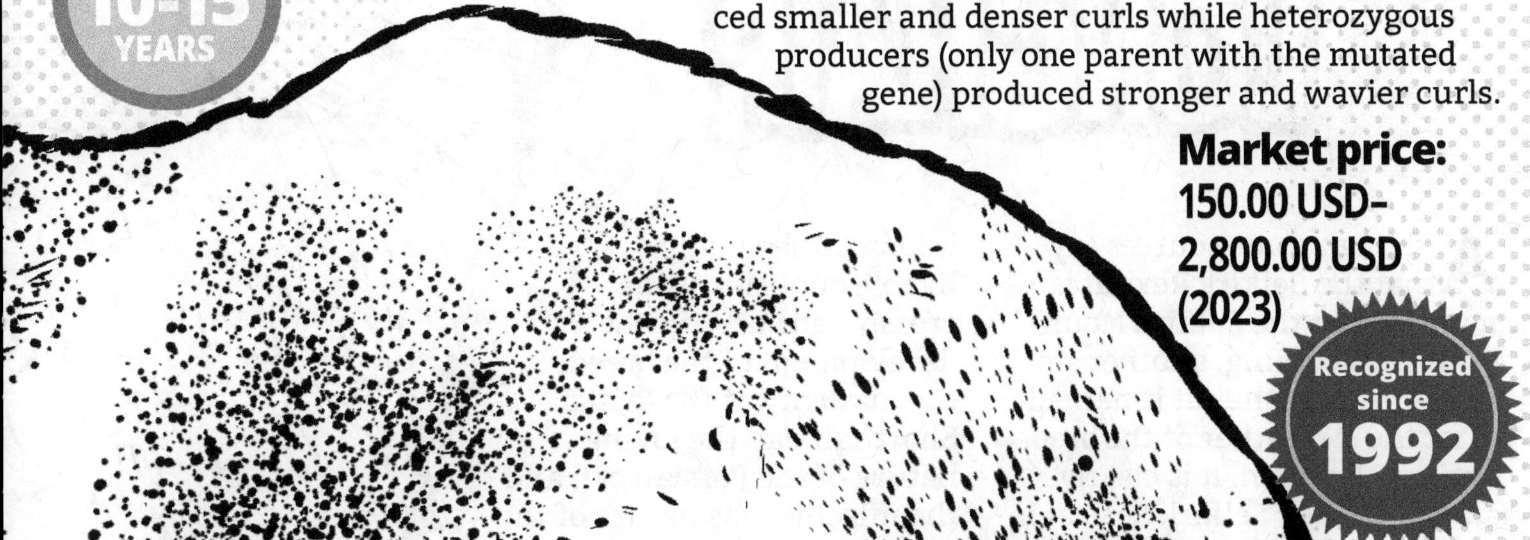

SELKIRK REX LONGHAIR

Although some sources say that the Selkirk Rex cat is named after the Selkirk Mountains in Wyoming, another story claims that the cat is named after the stepfather of the breeder Jeri Newman. It is clearly stated that the Selkirk Rex cat is probably the only pedigree cat named after a real person: the father-in-law and not the stepfather. To the appearance: The Selkirk Rex cat wears a lush and thick curly coat that is fluffier than that of the Cornish cat and Devon Rex cat and more splendid than that of the LaPerm cat. Of course, she has also inherited the many likable qualities of the crossbreeds used in the development of the breed: the maturity of the British Shorthair cat, the loving nature of the Persian cat, and the mischievous nature of the Exotic Shorthair cat. The quiet Selkirk Rex cat is very human-oriented. Therefore, she constantly follows her owner and wants body contact with him. Also, for this reason, she is sometimes affectionately called a dog cat.

Weight 6-12 lbs

ACF	ACFA	CAA	CCA-AFC	CCC	CCCA	CFA	FB	FIFe	GCCF	LOOF
✓	✓	✓	✓	✓	✓	✓	✓	✓	✓	✓

NZCF	SACC	TICA	WCF
✓	✓	✓	✓

Selection of the most important cat associations in the world: Australia, Canada, China, France, Germany, New Zealand, South Africa, UK and USA

Breeding start
1980s
United States

LIFE EXPECTATION
10-15
YEARS

Recognized since
1994

Market price:
150.00 USD–
2,800.00 USD
(2023)

REX CAT

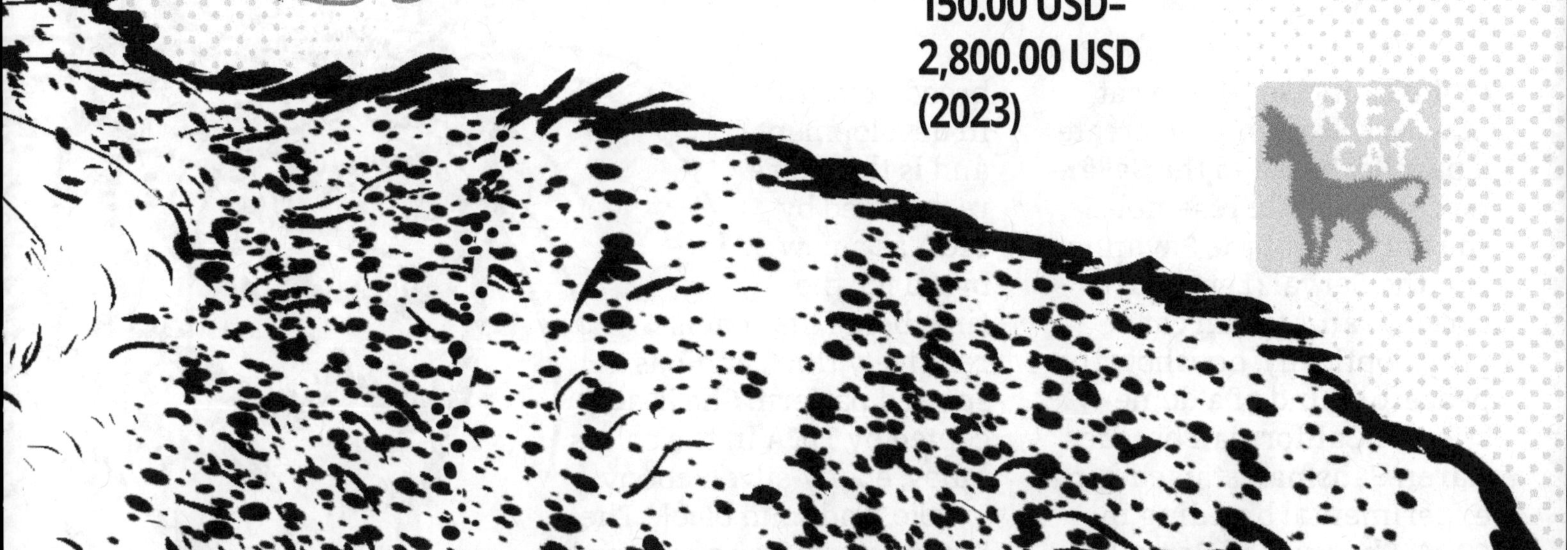

SERENGETI

The exotic Serengeti cat was the attempt to create a parallel breed to the Savanna cat. It should resemble the ancestor of the Savanna cat, the Serval (Leptailurus serval) - an African cat of prey - optically, but show the peaceful mind of a domestic cat. Californian breeder Karen Sausman started her experimental breeding in 1994. She crossed Bengal cats with spotted Oriental Shorthair cats. The result was a big cat with big ears and a dainty face. The appearance of the slender and muscular Serengeti cat resembles that of an Oriental Shorthair with longer legs and a longer neck. The temperamental Serengeti cat is a young cat breed. It is still in development and is therefore recognized by TICA as a new breed for the time being. Its coat is relatively thick and has a spotted pattern. She is accepted by TICA in the colors tabby, ebony-silver, ebony smoke, and solid black. The Serengeti cat has a relatively small nose but huge ears. The very active and playful cat has an alert gaze with interesting looking, attentive eyes that register everything.

BIG CAT

Breeding start
1990s
United States

ACF	ACFA	CAA	CCA-AFC	CCC	CCCA	CFA	FB	FIFe	GCCF	LOOF

NZCF	SACC	TICA	WCF
		✓	

Selection of the most important cat associations in the world: Australia, Canada, China, France, Germany, New Zealand, South Africa, UK and USA

ATELIER KAYMAK®

Science PAPERS

ORIGIN:

ASIA, UNITED STATES

Recognized since 2009

Market price:
600.00 USD–
2,800.00 USD
(2023)

LIFE EXPECTATION
10-15
YEARS

SEYCHELLOIS SHORTHAIR

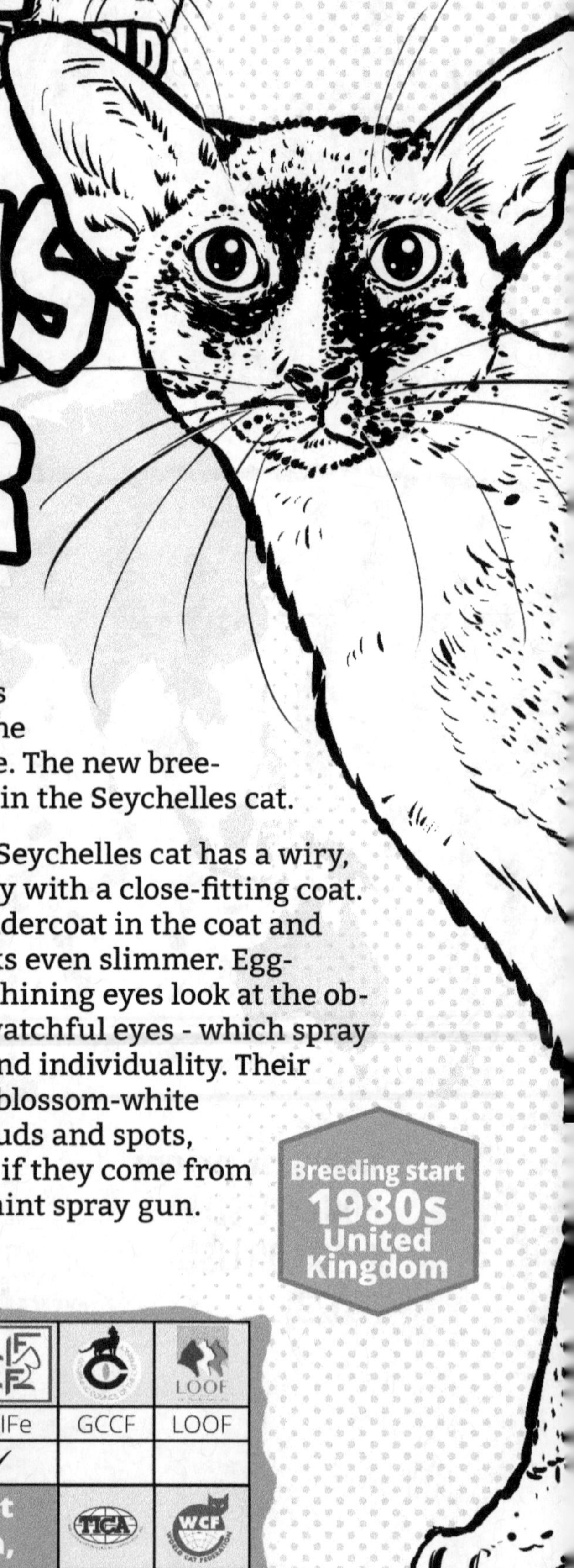

The Seychelles cat looks very similar to the Siamese cat but offers a unique coloring. The history of the Seychelles cat begins with the Siamese cat. Since the Siamese cat entered the western hemisphere, breeders tried to change its appearance. It has been crossed in countless experiments with other cats to breed more varieties in shape and color. This was also the case in the 1980s when the British breeder Patricia Turner dared to breed a new cat. Mrs. Turner used a Siamese cat and a bright, longhaired Persian cat as the genetic basis for a new variant. The offspring were then crossed with genes of the Oriental Shorthair cat - from which the Seychelles cat then got the elongated face. The new breeding resulted in the Seychelles cat.

A full-grown Seychelles cat has a wiry, muscular body with a close-fitting coat. She has no undercoat in the coat and therefore looks even slimmer. Egg-shaped, blue shining eyes look at the observer from watchful eyes - which spray intelligence and individuality. Their fur is light to blossom-white with color clouds and spots, which look as if they come from a disguised paint spray gun.

Breeding start
1980s
United Kingdom

ACF	ACFA	CAA	CCA-AFC	CCC	CCCA	CFA	FB	FIFe	GCCF	LOOF
							✓	✓		

NZCF	SACC	TICA	WCF
✓	✓	✓	✓

Selection of the most important cat associations in the world: Australia, Canada, China, France, Germany, New Zealand, South Africa, UK and USA

ORIGIN:

ASIA, UNITED KINGDOM

Perhaps the French have always had a little more time for art - because the Seychelles cat was recognized in 2005 only by the French FIFe (and thus also by the English branch Felis Britannica) as a new, own cat breed. For the evaluation of the Seychelles cat it is important, where exactly the colours sit. Therefore the FIFe jury divides the Seychelles cat into three varieties: Septieme - coloured dots above each part of the body; Huitieme - coloured parts on the head, legs and tail; Neuvieme - colouring only on the tail and head.

Recognized since 2005

LIFE EXPECTATION 12-15 YEARS

Market price:
500,00 USD

SEYCHELLOIS LONGHAIR

Two-colored Balinese cats are known as Seychellois cats or simply Seychelles cats. Breeder Patricia Turner was inspired, it is said, by reports of the ancestors of these beautiful cats living in the Seychelles archipelago. Seychelles cats are generally very temperamental and comparable to Siamese and Oriental cats. The ears of newborn Seychelles cats are so huge in relation to their body that some call them elephant cat babies :-). The cat associations use a color system in the classification to evaluate the exact white degree of Seychelles cats: 1 corresponds to solid black, and ten corresponds to solid white. That's less than 50 Shades of Grey- but enough to work with.

Breeding start
1980s
United Kingdom

LIFE EXPECTATION
12-15
YEARS

ACF	ACFA	CAA	CCA-AFC	CCC	CCCA	CFA	FB	FIFe	GCCF	LOOF
							✓	✓		

NZCF	SACC	TICA	WCF

Selection of the most important cat associations in the world: Australia, Canada, China, France, Germany, New Zealand, South Africa, UK and USA

ATELIER KAYMAK
Science PAPER
ORIGIN:
ASIA, UNITED KINGDOM
Recognized since 2005
Market price:
500.00 USD–
800.00 USD
(2023)
Weight

SHIH TZU

There are two controversial theories about the origins of the breed: One says that Tibetan monks developed Shi Tzu's cats as toys. The other claims that Tibetan monks managed to stealthily steal one of these crazy animals from a yeti strolling by, giggling confusedly to himself. The Shih Tzu cat is an unusual and remarkable breed of cat. It is regularly thrown out of cat shows for reasons that are as yet unexplained. The reasons could be that she is not good with other cats, makes bad jokes. Or because she is impudent and licks everyone. Or she has incredibly bad taste in fashion. But what cats and award winners don't want to see is: that she is very loyal. Yes, she is loud. She can even bark loudly when she doesn't like something. Sometimes she barks for no reason at all because she enjoys it. The Shih Tzu cat is unfortunately still ignored by all cat associations. In Germany, the Animal Welfare Act, according to § 11b (Qualzucht-Paragraph), prohibits the breeding of the Shih Tzu.

Breeding start 1600s China

LIFE EXPECTATION 10-18 YEARS

ANCIENT ORIGIN

ACF	ACFA	CAA	CCA-AFC	CCC	CCCA	CFA	FB	FIFe	GCCF	LOOF
☺	☺	☺	☺	☺	☺	☺	☺	☺	☺	☺

NZCF	SACC	TICA	WCF
☺	☺	☺	☺

Selection of the most important cat associations in the world: Australia, Canada, China, France, Germany, New Zealand, South Africa, UK and USA

ORIGIN:

TIBETAN HIGHLANDS, CHINA

Market price:
500.00 USD–
3,000.00 USD
(2023)

SHORT HAIR

SIAMESE

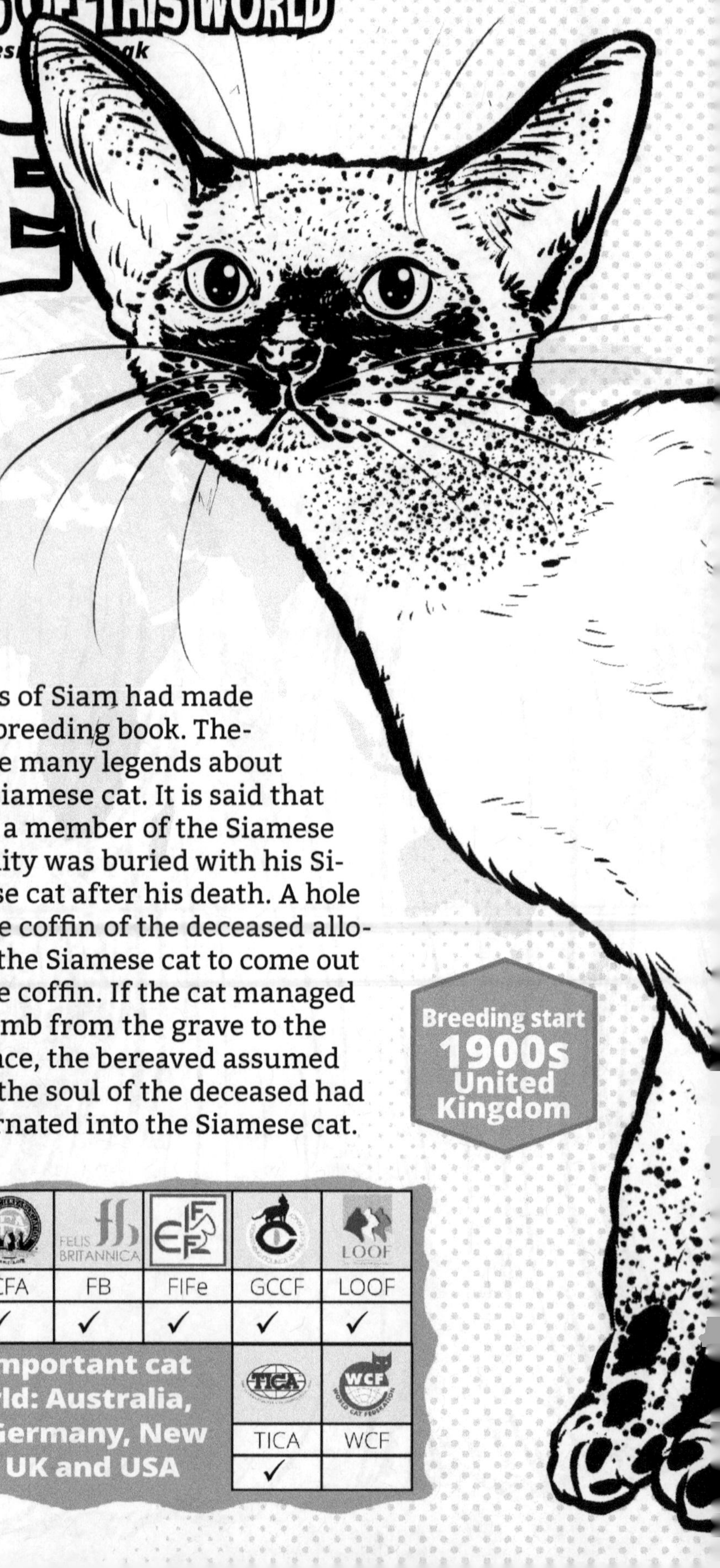

The Siamese cat is one of the oldest and best-known cat breeds in the world. It comes from the old Siamese - today's Thailand in Southeast Asia. The Siamese cat has long served as an inexhaustible source for many cat owners who have been involved in the development of new cat breeds since the foundation of cat associations. The Siamese cat is a shining example of cat-like elegance and is the perfect benchmark for cat breeders worldwide. It is so exceptional that it has been exhibited in Europe Zoos. This is because the cat is a partial albino and has blue eyes due to a mutation. Another distinctive feature of the Siamese cat is its bright white, silky coat - which runs dark in so-called points (face, ears, tail, and paws). She is also the most famous of all 17 listed royal cats in the famous Tamra Maew - the book about cats - which once the Kings of Siam had made as a breeding book. There are many legends about the Siamese cat. It is said that once a member of the Siamese nobility was buried with his Siamese cat after his death. A hole in the coffin of the deceased allowed the Siamese cat to come out of the coffin. If the cat managed to climb from the grave to the surface, the bereaved assumed that the soul of the deceased had incarnated into the Siamese cat.

Breeding start
1900s
United Kingdom

ACF	ACFA	CAA	CCA-AFC	CCC	CCCA	CFA	FB	FIFe	GCCF	LOOF
✓	✓	✓	✓	✓	✓	✓	✓	✓	✓	✓

NZCF	SACC	TICA	WCF
✓	✓	✓	

Selection of the most important cat associations in the world: Australia, Canada, China, France, Germany, New Zealand, South Africa, UK and USA

ATELIER KAYMAK®

Science

PAPERS

The first Siamese cats in the western hemisphere were presented in 1885 at a cat exhibition in London's Crystal Palace. At that time, the Siamese cat was still called the Royal cat of Siam. 1892 the first breed standards were created. Around the turn of the century, the Siamese cat was already bred in England, France, and the USA. 1901 the first English Siamese cat Club was founded.

Market price:
600.00 USD–3,000.00 USD (2025)

ANCIENT ORIGIN

Recognized since 1878

TOP 10

Gangliosidosis, Progressive Retinal atrophy (PTR), Intestinal Diseases (IBD), Diseases of eyes, ears and teeth, Cancer, Obsessive-compulsive disorder (OCD)

LIFE EXPECTATION 15-20 YEARS

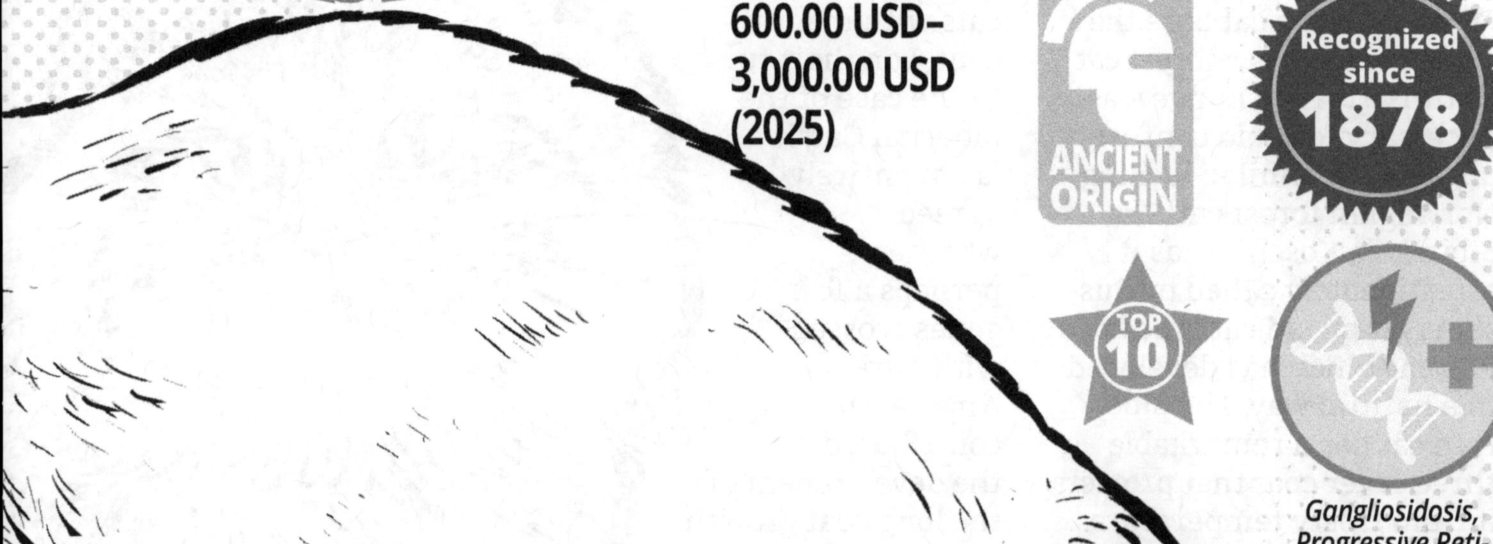

SIBERIAN CAT

The Siberian cat bore the name Siberian forest cat until 1991. The change was necessary to avoid confusion with the similar-looking Norwegian forest cat. The Sibirskaja Koschka - as it is affectionately called by Russians - is an old cat breed of Russia, which has developed in a natural way. The Siberian cat has a remarkable three-layer coat that protects it from frosty temperatures. A layer of dense undercoat regulates the heat; the Grannen hair acts as an insulating layer, and the covering coat protects from sun rays and is water repellent. The Siberian cat belongs together with the Norwegian Forest Cat and the Maine Coon cat to the group of the so-called Forest cats: the big, strong cats with magnificent, Longhair fur. In the case of the Siberian Cat, it is not entirely agreed whether perhaps a few genes crossed with Turkish Angora cats contributed to the development of the long coat growth. It would be conceivable because, over the old trade routes of the Silk Road, many of these cats reached so some places. There are, for example, historical travel descriptions from the 19th century, in which one speaks of conspicuously longhaired Cumanese cats from the

ACF	ACFA	CAA	CCA-AFC	CCC	CCCA	CFA	FB	FIFe	GCCF	LOOF
✓	✓	✓	✓	✓	✓	✓	✓	✓	✓	✓

NZCF	SACC	TICA	WCF
✓	✓	✓	

Selection of the most important cat associations in the world: Australia, Canada, China, France, Germany, New Zealand, South Africa, UK and USA

Caucasus. In the early 1980s, some of the resilient cats accompanied Russian construction workers who laid a section of the oil pipeline in what was then the East German Democratic Republic (GDR) within the framework of the Council for Mutual Economic Assistance (RGW) - the Eastern Bloc's response to the US Marshall Plan. The pipeline, called Druschba-Trasse by East Germans, began at the western foothills of the Ural Mountains and ran for 2,760 km through Bulgaria, Czechoslovakia, and Poland in the GDR. The cats of the Russian construction workers were therefore called Trassenkatzen, line-cats. East German breeders were now interested in her and started a breeding program in 1987. The first Siberian forest cat registered in their official studbook bore the poetic name Wuschel (fuzzy-wuzzy). Still, in the same year, the breeding of the Siberian forest cat began in the Soviet Union and in West Germany. Two years later, the Siberian forest cat finally arrived in the USA. In 1991 the name was changed, and one year later, in 1992, the Siberian cat was recognized by the WCF as a new cat breed.

Market price:
1,500.00 USD–7,000.00 USD (2025)

Breeding start 1980s East Germany

Recognized since 1882

(with pure white coat) Congenital hearing loss, deafness, disturbances of the sense of equilibrium, Eye tremor

LIFE EXPECTATION 14-16 YEARS

LUXUS

ATELIER KAYMAK'S CREATIVE- & STUDY BOOKS

ALL CAT BREEDS OF THIS WORLD

by Nuesret Kaymak

SHORT HAIR

Breeding start 1950s United States

SINGAPURA

The Singapura cat is one of the rarer race-cats. The cat with the small stature and the small physique has its origin in the street canyons of the metropolis of the same name. The Asian city-state of Singapore is the smallest state in Southeast Asia. The modern metropolis with its 5.6 million inhabitants is one of the wealthiest countries in the world and, along with Hong Kong, Asia's most important financial center. Singapore is about the size of Hamburg and is spread over several islands: one main island, three larger and 58 smaller islands. Despite the many people, the many cars on paved roads, and the skyscrapers made of glass and steel - the living space for a small cat has arisen here. The tiny animal weighs only two and a half to three kilograms and therefore fits with its dwarflike body

ACF	ACFA	CAA	CCA-AFC	CCC	CCCA	CFA	FB	FIFe	GCCF	LOOF
✓	✓	✓	✓	✓	✓	✓	✓	✓	✓	✓

NZCF	SACC	TICA	WCF
✓		✓	✓

Selection of the most important cat associations in the world: Australia, Canada, China, France, Germany, New Zealand, South Africa, UK and USA

Hypokalaemia (potassium deficiency), Progressive Retinal atrophy (PTR)

through almost all of the city's drainage pipes. In fact, the Singapura cat is also called Drain cat by the inhabitants. Her breeding history is as short as her body length: In 1974, Hal and Tommy Meadow noticed the little alley cat during their visit to the city. The couple took the kitten with them to the USA, where they started breeding the Singapore cat with her. In 1982 the Singapore cat received the championship status from the American TICA. The city-state of Singapore used the world-famous cat as a mascot in 1991 to boost tourism. In 1993 it was declared a national symbol of the city-state.

Market price:
1,200.00 USD–
3,000.00 USD
(2023)

Recognized since 1882

LIFE EXPECTATION 14-16 YEARS

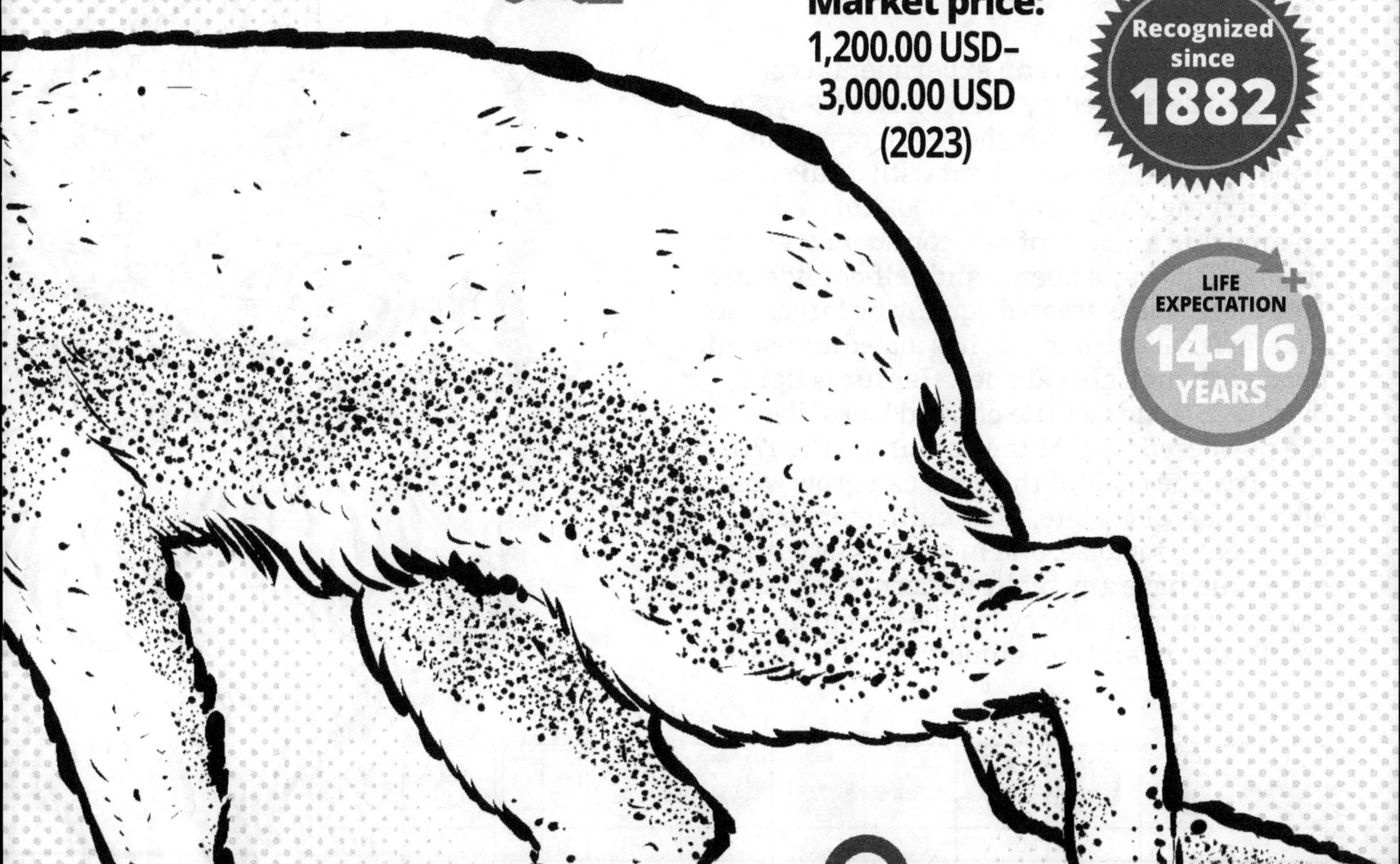

Breeding start
1990s
United States

SKOOKUM SHORTHAIR

The Skookum cat is an experimental cat breed developed by the American Roy Galusha in the 1990s. It is the result of crossing a short-legged Munchkin cat with a LaPerm cat carrying curly fur. The Skookum cat has an amazing amount of self-confidence, which she displays quite openly and self-confidently. Therefore, she is treated equally by larger cats that tower above her. She has a medium-sized head with high cheekbones. Her fur is tightly waved, stands a little off, and looks like the powdered wigs of the nobility in the 17th century. The color of their fur can grow white, black, blue, chocolate, lilac, cinnamon, fawn, red, cream, tortoise, and in variations with silver and have any tabby pattern. The playful Skookum cat is very mobile despite her small stature and can climb trees effortlessly.

ACF	ACFA	CAA	CCA-AFC	CCC	CCCA	CFA	FB	FIFe	GCCF	LOOF

NZCF	SACC	TICA	WCF
		E	

Selection of the most important cat associations in the world: Australia, Canada, China, France, Germany, New Zealand, South Africa, UK and USA

The cat, which combines the kindness of a Munchkin cat and the curiosity of a LaPerm cat, is available in many different colors. The word Skookum is a slang used in the Pacific Northwest to express wildness, size, and strength, giving the breed a breeze of delicious irony. Fun Fact: a Munchkin- LaPerm litter can contain Munchkin, Skookum, and LaPerm newborns! The Skookum cat has not yet been recognized by any association but has been approved by TICA for registration as an experimental variety.

Dwarf Cat

DESIGNERCAT

LIFE EXPECTATION 10-15 YEARS

Market price:
500.00 USD–
2,000.00 USD
(2023)

Recognized since 1882

Weight 3-7 lbs

SNOWSHOE

The most famous cat of the social media age – The Grumpy cat - was a Snowshoe cat. This cat breed is essentially a Siamese cat with a point drawing - a white part of paws, face, and chin. The breed was a Siamese cat and a two-tone American Shorthair cat. The American Dorothy Hinds-Deugherty bred Siamese cats in Philadelphia in the 1960s and found three kittens with white points in one of the litters of her breeding cats. Usually, the color combination was precisely the other way around, so the kittens with the white socks stood out and were called Snowshoes. She crossed the adult cats with American short-haired cats and the resulting Offspring again with Siamese cats - which had the consequence that the coat pattern could be consolidated and passed on.

Breeding start
1960s
United States

ACF	ACFA	CAA	CCA-AFC	CCC	CCCA	CFA	FB	FIFe	GCCF	LOOF
✓	✓	✓			✓		✓	✓	✓	✓

NZCF	SACC	TICA	WCF
		✓	

Selection of the most important cat associations in the world: Australia, Canada, China, France, Germany, New Zealand, South Africa, UK and USA

In practice, however, breeding proves to be very laborious due to the inconsistency in achieving the desired color pattern. This means that there are very few Snowshoe cats in the world, even though the breed has existed for many decades. In 1982 Snowshoe was awarded championship status by TICA and ACFA.

Market price:
2,000.00 USD–
4,000.00 USD
(2023)

Recognized since 1983

LIFE EXPECTATION 12-15 YEARS

LUXUS

SHORT HAIR

Breeding start
1950s
United States

SOKOKE

The Sokoke cat or Sokoke forest cat was known for a long time under the name Khadzonzo wild cat, a cat species native to the Kenyan Arabuko-Sokoke forest in East Africa. The Sokoke forest consists of three types of forest and is known for its plants and animals, which otherwise only exist on the island of Zanzibar. In the middle of the park flows the Athi-Galana-Sabaki, a 700 km Long River that joins the Indian Ocean in the west. The river became famous through the Danish writer Karen Blixen, who lived in the area at the beginning of the 20th century as a coffee farmer and became famous worldwide with her novel Out of Africa. Many decades later, in the 1970s, the animal painter Jeni Slater, probably inspired by Blixen - who also painted - lived with her husband close to the Sokoke forest. And with Jeni Slater begins the modern breeding history of the Sokoke cat.

ACF	ACFA	CAA	CCA-AFC	CCC	CCCA	CFA	FB	FIFe	GCCF	LOOF
			✓				✓	✓		✓

NZCF	SACC	TICA	WCF
		✓	✓

Selection of the most important cat associations in the world: Australia, Canada, China, France, Germany, New Zealand, South Africa, UK and USA

Weight
7-13
lbs

During one of her excursions, she discovered a mother cat with her kittens at the edge of the forest, which was then threatened by massive deforestation. The cat had hidden with her litter behind the freestanding roots of a big tree. The amiable cat had large eyes and ears and the coat pattern of a wild cat. Knowing of the decline in the population, Mrs. Slater did not think long and took the two youngest kittens home with her. The cats grew up and prospered into magnificent animals. They formed the basis for Slater's breeding of the Sokoke forest cats, as she called them. In 1983 Gloria Moldrup from Denmark traveled to Kenya (probably also inspired by Karen Blixen) and visited Slater and her cats. The Dane - herself a cat lover - immediately fell in love with the cats, and Slater gave her a pair. At home in Denmark, Moldrup immediately started to breed the cats. Already in 1984, she introduced the Sokoke Forest Cat to an interested public in Copenhagen. FIFe recognized the breed in 1993.

SOMALI

Breeding start
1960s
United Kingdom

The Somali cat is a variant of the Abyssinian cat - and in comparison to this, it has a longer and softer coat on the chest, chin, legs, and tail. Long hair variants of the Abyssinian cat repeatedly appeared - but were always evaluated as unsuccessful breeding results and not followed up. Abyssinian cats were subject to strict breeding regulations, and every deviation was considered a mortal sin. But there are probably some cats that have attracted attention: In 1953, a long-haired Abyssinian cat named Raby Chuffa of Selene made headlines in the USA; in 1963, the Canadian breeder Mary Mailing presented her sensational long-haired Abyssinian cat. At the end of the 1960s - everything became freer: love as well as breeding requirements - the long-haired Abyssinian gradually gained friends and became popular. It even got the nickname fox cat. This promising circumstance made cat breeders in all parts of the world give in. From now on, they began to breed long-haired Abyssinian cats. However, the American breeder Evelyn Mague caused the actual popularity boost: in order to make the longhaired cat better known, she gave it the exotic name Somali. So she did not only make the cat more interesting (everyone immediately thought of Somalia, Africa, and of course: Lions!) but she also clearly separated it from the Abyssinian cat. In 1979 the Somali cat was recognized by the CFA as a new cat breed.

LIFE EXPECTATION
10-12
YEARS

ACF	ACFA	CAA	CCA-AFC	CCC	CCCA	CFA	FB	FIFe	GCCF	LOOF
✓	✓	✓	✓	✓	✓	✓	✓	✓	✓	✓

NZCF	SACC	TICA	WCF
✓	✓	✓	

Selection of the most important cat associations in the world: Australia, Canada, China, France, Germany, New Zealand, South Africa, UK and USA

ATELIER KAYMAK
Science
PAPERS

Market price:
1,000.00 USD–4,500.00 USD
(2023)

Recognized since 1979

Pyruvate kinase deficiency (PK)

SPHYNX

ALIAS CANADIAN SPHYNX

The first Sphynx cat was a cat named Prune and was born in 1966 in Toronto of a black and white domestic cat. Prune was born as a hairless cat. As with all Rex cats, it was later discovered that the hairless Sphynx gene affects not only hair growth but also the body shape massively. Prune's mother was described as a normal-sized medium build cat, while her hairless son was very lanky and slender. The local breeder Yania Bawa, who specialized in Siamese cats, acquired Prune and his mother and started a new breeding program specially designed for Sphynx cats. Although the Sphynx breed is more or less accepted today, it was very controversial in its beginnings. Although the championship title won in 1971, questions about health problems remained. There were also heated discussions about the appearance of the Sphynx cat, which finally led to the loss of the title and the subsequent disappearance of this cat breed. However, the Sphynx breeding program was continued in other parts of Europe. As a result, the Sphynx cat regained its status from TICA in 1990. Since it has no coat, the Sphynx cat needs an intensive care program to avoid excessive exposure to cold and sunlight.

Breeding start
1868
United Kingdom

ACF	ACFA	CAA	CCA-AFC	CCC	CCCA	CFA	FB	FIFe	GCCF	LOOF
✓	✓	✓	✓	✓	✓	✓	✓	✓	✓	✓

NZCF	SACC	TICA	WCF
✓		✓	

Selection of the most important cat associations in the world: Australia, Canada, China, France, Germany, New Zealand, South Africa, UK and USA

DESIGNERCAT

ATELIER KAYMAK

Science

PAPERS

The Sphynx cat has an amiable nature and is a loyal, loving companion. In countries like Germany, the Animal Protection Act prohibits the breeding of the Sphynx-cat.

Market price:
2,500.00 USD–12,000.00 USD (2025)

HAIRLESS CAT

Recognized since 1882

LUXUS

LIFE EXPECTATION 14-16 YEARS

Myopathy (muscle weakness), Hypokalaemia (potassium deficiency)

TOP 10

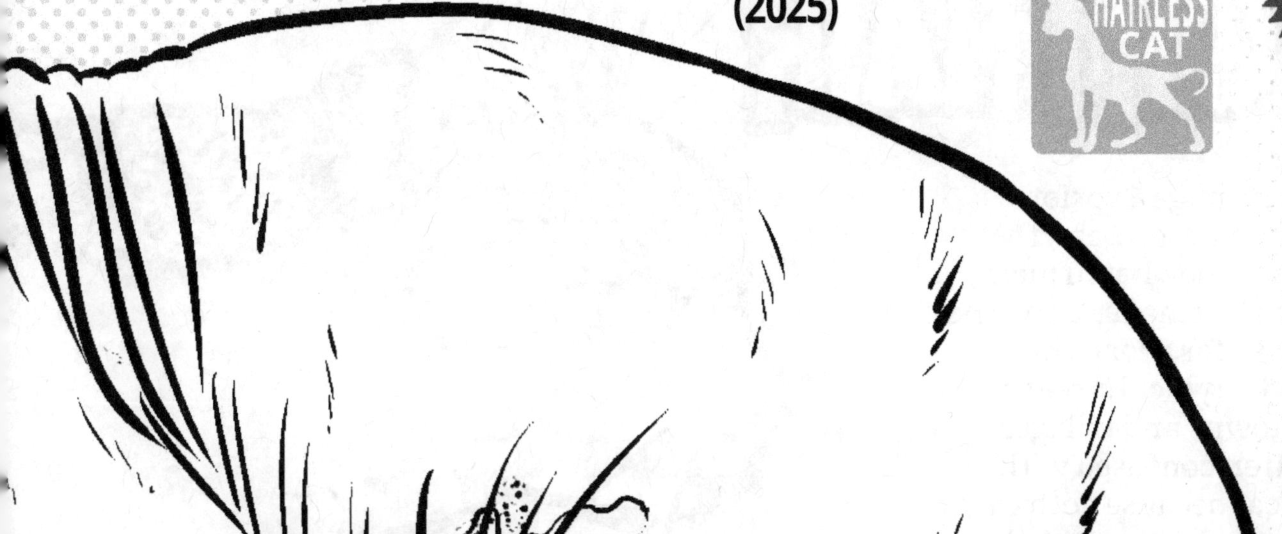

ST HELEN CAT SHORTHAIR

In contrast to its huge Cypriot neighbor, the Aphrodite cat, the St Helena cat is somewhat firmer, fuller, and also has smaller ears. The St Helena cat was for a long time only considered a typical street cat of the Cypriot towns and villages. It was and is often confused with the Aphrodite cat because both cat breeds have just recently started a breeding history. Among the cat associations, there is indeed some confusion about the individual status of the cats. The long-time comfortably living cats were not noticed for many millennia and tolerated as street cats. In the context of a society that is becoming increasingly aware of environmental and animal protection issues, many of the inhabitants also became interested in the cats living in their surroundings. This led naturally also to the question around the origin of the cats.

ACF	ACFA	CAA	CCA-AFC	CCC	CCCA	CFA	FB	FIFe	GCCF	LOOF

NZCF	SACC	TICA	WCF

Selection of the most important cat associations in the world: Australia, Canada, China, France, Germany, New Zealand, South Africa, UK and USA

ORIGIN:

CYPRUS, MIDDLE EAST, EGYPT

The knowledge of the antiquity research made the inhabitants proud of the history of their cats, which were treated now much more respectfully. The St Helena Shorthair cat can also be found outside Cyprus on numerous Greek islands. The St Helena cat has so far only been recognized by the local Cyprus Feline Society.

Market price:
300.00 USD–600.00 USD (2023)

Breeding start
2000s
Cyprus

Recognized since
2009

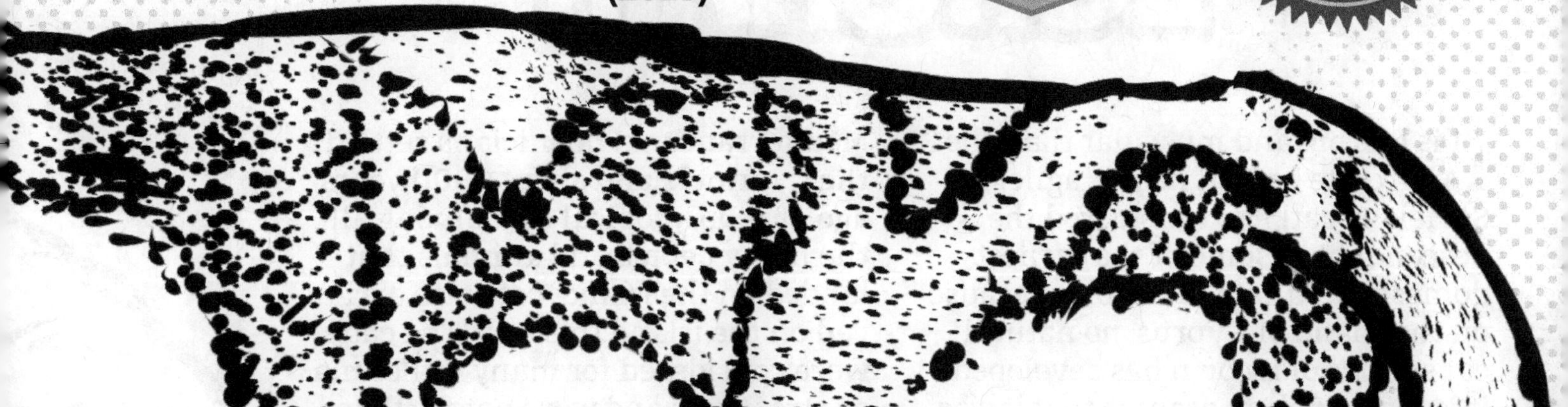

LIFE EXPECTATION
12-15
YEARS

ST HELEN CAT LONGHAIR

Breeding start
2000s
Cyprus

It is less big and muscular than the Aphrodite cat but no less agile: the St Helena cat! It is considered the descendant of some of the earliest domesticated cats in history. Because on the island of Cyprus, no natural cat species of its own has developed - it is reasonable to assume that it was brought to the island by humans. This is credible because 12,000 years ago, people settled in the immediate vicinity of the island. Early Mesopotamian cultures such as Sumer, the Akkad Empire, the Assyrian Empire, and Phoenicia expanded as well as ancient Egypt and the Roman Empire. Countless scouts, settlers, and merchants landed on the islands of the Mediterranean: Phoenicians settled the Cypriot island about 9000 years ago. It is only logical that among all the many people and animals that landed on the island, cats also arrived on the island this way. The cats were considered for many millennia as street cats and were not further noticed. In the context of a society increasingly sensitizing itself for environmental and animal protection - many inhabitants began to be interested also in their cats. This led to the question about the origin of the cats. The findings of the antiquity research made

ACF	ACFA	CAA	CCA-AFC	CCC	CCCA	CFA	FB	FIFe	GCCF	LO

NZCF	SACC	TICA	WCF

Selection of the most important cat associations in the world: Australia, Canada, China, France, Germany, New Zealand, South Africa, UK and USA

ORIGIN: CYPRUS, MIDDLE EAST, EGYPT

the inhabitants proud of the history of their cats, which were now treated much more respectfully. The St Helena long-haired cat can also be found outside Cyprus on numerous Greek islands. The St Helena longhaired cat has only been recognized by the local Cyprus Feline Society.

Recognized since 2009

Market price:
300.00 USD–
600.00 USD
(2023)

LIFE EXPECTATION 12-15 YEARS

SUPHALAK

The Suphalak cat bears the name of a very old cat breed related to the modern Burmese cat of today: the Suphalak or Thuong Daeng cat is listed in the Tamra Meaw (the Meauw actually corresponds to the Miao sound of a cat) - the textbook for cats, the studbook for the royal court cats of old Siam. It appears in fourth place and is translated as Excellency. One of the poems in Tamra Meaw says that the Suphalak is as rare as gold, - which describes the rarity value of this cat. The descendant of such a cat, awarded with majestic honors, is recognizable by its short reddish-brown, copper-colored coat. The cat breed was considered lost for a long time. Since the 2010s, breeders like Kamnan Preecha Pukkabut have been working in Thailand to breed Suphalak cats with their original appearance. Since 2014, there has been an association that has joined as a national heritage to preserve the Suphalak - The International Maew Boran Association (TIMBA).
The Suphalak cat was recognized by the American cat Association in 2015 (not in the table).

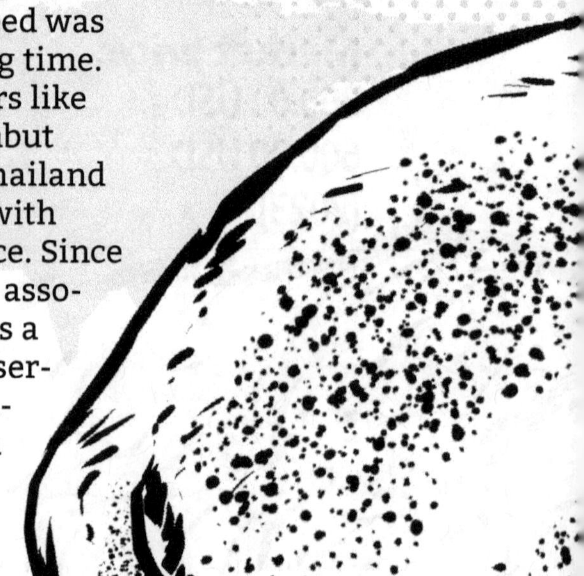

Breeding start
2010s
Thailand

LIFE EXPECTATION
12,15
YEARS

ACF	ACFA	CAA	CCA-AFC	CCC	CCCA	CFA	FB	FIFe	GCCF	LOOF

NZCF	SACC

Selection of the most important cat associations in the world: Australia, Canada, China, France, Germany, New Zealand, South Africa, UK and USA

TICA	WCF

ATELIER KAYMAK
Science
PAPERS
269
ORIGIN:
SOUTHEAST ASIA, THAILAND
Recognized since 2009
ANCIENT ORIGIN
Market price:
1,200.00 USD–
1,500.00 USD
(2023)

ATELIER KAYMAK'S CREATIVE- & STUDY BOOKS

ALL CAT BREEDS OF THIS WORLD

by Nuesret Kaymak

SHORT HAIR

TIBETAN

The Tibetan cat is still a relatively young cat breed. It was first developed in 1992 by the Dutch breeder Agnes Driessen and originated from the crossing of a Balinese cat - a semi-longhaired Siamese cat - and a Burmese cat. The Tibetan cat is essentially a semi-long-haired Tonkinese cat and is often mistaken for this cat. But the striking differences are the long coat and the large tail, which is the special characteristic of this cat. The Tibetan cat is also often confused with the Asian Manul (Otocolobus manul), a small predatory cat that is related to cheetah and lynx because of its name and is popularly known as the Tibetan cat. The character trait of the playful Tibetan cat can be well described in one sentence:

she is convinced that people were only brought to earth to love Tibetan cats.

The fur is available either in mink, point, or sepia. The Tibetan cat is still in the early stages of recognition by GCCF and TICA. So far, it has only been recognized by Dutch associations.

ACF	ACFA	CAA	CCA-AFC	CCC	CCCA	CFA	FB	FIFe	GCCF	LOOF

NZCF	SACC	TICA	WCF
		A	

Selection of the most important cat associations in the world: Australia, Canada, China, France, Germany, New Zealand, South Africa, UK and USA

ORIGIN:
SOUTHEAST ASIA, NETHERLANDS
Recognized since 1997
Breeding start 1992 Netherlands
LIFE EXPECTATION 13-17 YEARS
Market price:
650.00 USD–
1,000.00 USD

SHORT HAIR

Breeding start 1980s United States

THAI ALSO WICHIEN MAAT

The Thai cat or Wichien Maat originated in the 1970s. At that time, there was a trend to revive the Siamese cat's original, slim appearance. Therefore, the Thai cat was called Old-Style Siamese in the English world. In the 1990s, paradoxically, more and more Burmese cats - but also Rex and Colorpoint cats - were crossbred. The Thai cat is very human-oriented - which is why one speaks of it as a dog cat. It has a head rounder than the modern Siamese cat and mild facial features. Her almond-shaped eyes rise slightly on the outside, giving the cat a curious, cheeky expression. In 1993 the WCF accepted the Old-Style Siamese as a new breed. At the beginning of 2000, the Old-Style Siamese Club (OSSC) was founded in the UK, dedicated to breeding the traditional appearance of Siamese cats. In 2001 some Wichien-Maat cats were imported directly from Thailand to complete the breeding program. In 2007 the cat breed received its own competition class under the name Thai cat from TICA.

LIFE EXPECTATION 12-16 YEARS

ACF	ACFA	CAA	CCA-AFC	CCC	CCCA	CFA	FB	FIFe	GCCF	LOOF
							✓	✓	✓	✓

NZCF	SACC	TICA	WCF
		✓	✓

Selection of the most important cat associations in the world: Australia, Canada, China, France, Germany, New Zealand, South Africa, UK and USA

ATELIER KAYMAK
Science
PAPERS
273
ORIGIN:
SOUTHEAST ASIA, USA
Market price:
500.00 USD–
2,000.00 USD
(2023)
Recognized
since
2007
Weight
8-15
lbs

TIFFANIE

The Tiffanie cat was first bred in the 1980s by Jeane Bryson in England. The Tiffanie cat - which should not be confused with the American Tiffany cat - is essentially a long-haired Burmese cat. It differs from this only by its long silky fur, the ruff, and the bushy tail. Although it did not comply with the breeding regulations, a number of European breeders accidentally bred long-haired versions of the Burmese cat for some time. Because they were not accepted as purebred cats - the unwanted longhaired cats were often neutered and sold as regular domestic cats. Some specimens had fortunately been kept, and these formed the foundation for a new breed of cat, initially known as the Asian Longhair cat. The breed was awarded championship status by the GCCF in 2003.

Breeding start
1980s
United Kingdom

ACF	ACFA	CAA	CCA-AFC	CCC	CCCA	CFA	FB	FIFe	GCCF	LOOF
									✓	

NZCF	SACC	TICA	WCF
✓	✓		✓

Selection of the most important cat associations in the world: Australia, Canada, China, France, Germany, New Zealand, South Africa, UK and USA

ATELIER KAYMAK

ORIGIN:

SOUTHEAST ASIA, UK

LIFE EXPECTATION 7-12 YEARS

Market price: 300.00 USD– 1,000.00 USD (2023)

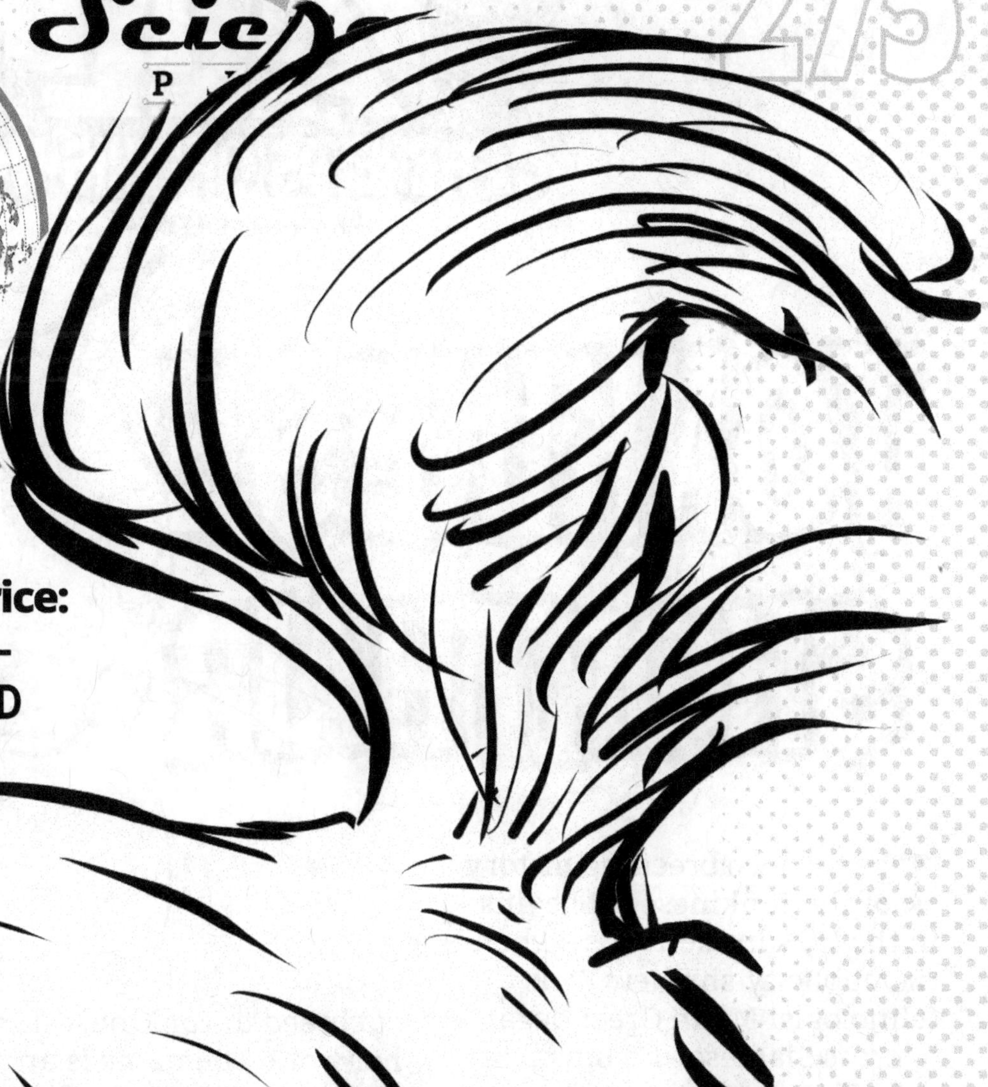

Recognized since

TONKINESE SHORTHAIR

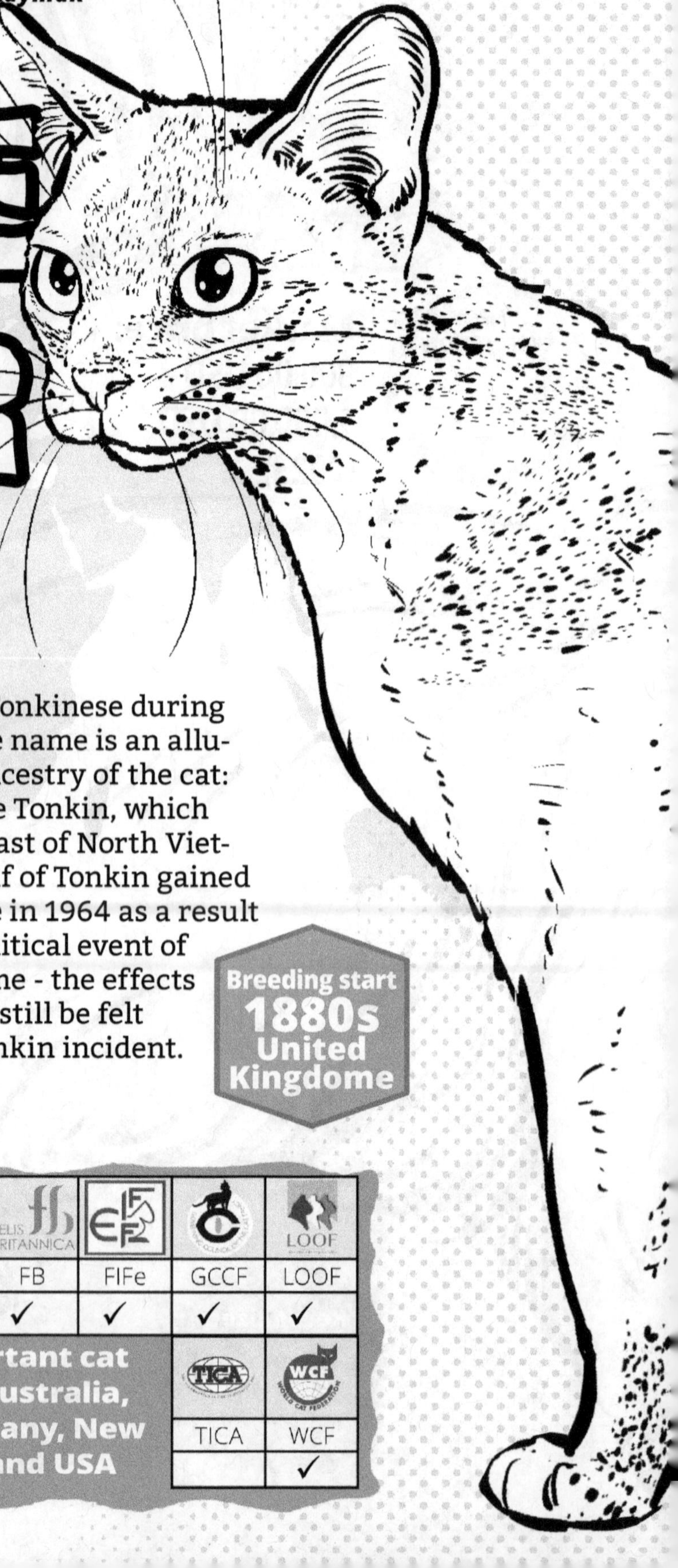

The modern breeding history of the Tonkinese cat begins in the USA in the 1950s - when the publicity-shy New York felinologist Milan Greer began to cross Siamese and Burmese cats. Since he kept his breeding results to himself, there was a movement in the matter only a decade later. In the 1960s, Canadian Margaret Conroy and American Jane Barletta continued breeding the Tonkinese cat. Conroy worked on a so-called Golden Siamese cat and changed the name of the new cat breed to Tonkinese during her work. The name is an allusion to the ancestry of the cat: the landscape Tonkin, which lies on the coast of North Vietnam. The Gulf of Tonkin gained dubious fame in 1964 as a result of a world political event of the same name - the effects of which can still be felt today: the Tonkin incident.

Breeding start
1880s
United Kingdome

ACF	ACFA	CAA	CCA-AFC	CCC	CCCA	CFA	FB	FIFe	GCCF	LOOF
✓	✓	✓	✓			✓	✓	✓	✓	✓

NZCF	SACC	TICA	WCF
✓	✓		✓

Selection of the most important cat associations in the world: Australia, Canada, China, France, Germany, New Zealand, South Africa, UK and USA

The alleged attack of North Vietnamese gunboats on US warships, which led to the Tonkin Resolution and the subsequent entry of the USA into the Vietnam War. Thus, the pretty Tonkinese cat unintentionally recalls the event that has forever burned itself into the collective memory of humanity. In 2001, the CFA shifted the Tonkinese cat from the variant class to its own fully recognized champion class.

Recognized since 1984

LIFE EXPECTATION 10-16 YEARS

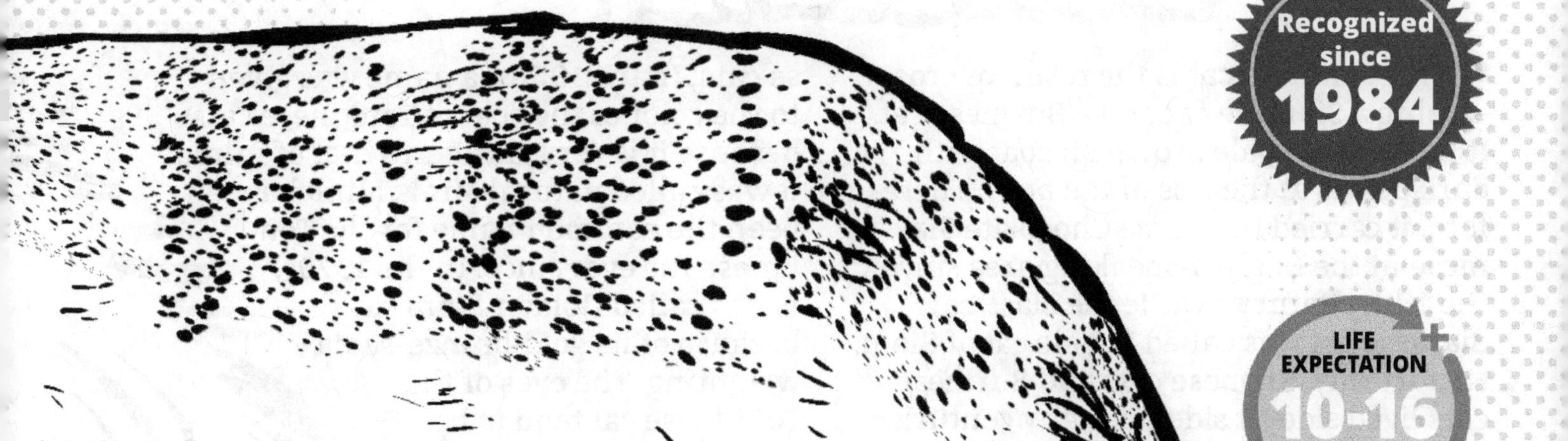

Market price:
400.00 USD–
1,500.00 USD
(2023)

LONG HAIR

TONKINESE LONGHAIR

ANCIENT ORIGIN

The Tonkinese cat is the result of crossing a Siamese cat and a Burmese cat. It has a short, pale brownish coat with dark points at the tips of the body. Therefore, it carried the name Chocolate Siamese in the English-speaking area since the 18th century - while the classical Siamese cat was called Royale cat of Siam. Historically, Siamese cats and Burmese cats lived side by side for many centuries in Southeast Asia and, in all probability, mated. This is supported by illustrations in old manuscripts from the 14th century, which depict the same pale brownish cats that are confusingly similar to Tonkinese cats. In the 1960s, a name was given to the color of the coat of Tonkinese cats: because it is very similar to that of mink, it was called natural Mink. Mink has been the common name for the Tonkinese fur ever since. In the 1970s, it was called Golden-Siamese because of its gold-bronze-sepia weighting. The eyes of the Tonkinese cat tend to be slightly greener than those of the Siamese cat. Except FIFe, all cat federations today recognize the Tonkinese cat as a new breed.

Breeding start 1880s United Kingdom

LIFE EXPECTATION 10-16 YEARS

Market price:
400.00 USD–1,500.00 USD (2023)

ACF	ACFA	CAA	CCA-AFC	CCC	CCCA	CFA	FB	FIFe	GCCF	LOOF

NZCF	SACC	TICA	WCF

Selection of the most important cat associations in the world: Australia, Canada, China, France, Germany, New Zealand, South Africa, UK and USA

ATELIER KAYMAK
Science
PAPERS
279
Recognized since 1984
ORIGIN:
SOUTHEAST ASIA, UK
Hypokalaemia (po-
tassium deficiency)

SHORT HAIR

Breeding start
1980s
Soviet Union

TOYBOB

The tiny Toybob cat is often referred to as the smallest cat in the world. The extremely rare cat with a stumpy tail was developed in the 1980s in the Russian regions of Rostov and Ural. It is assumed that both the small stature and the stumpy tail were spontaneous mutations of Russian wild cats. Despite its compact size, the Toybob cat is muscular - but weighs just three kilograms when fully grown. The Toybob cat is bred as a short-haired or semi-longhaired cat. It has remarkable, large, and expressive eyes, which stand out clearly in comparison to the other characteristics of the cat. Toybob cats like to be hugged and carried around. Therefore, they are called the Chihuahuas of cats by evil tongues. TICA classified them as an experimental new breed in 2009 and as a temporary new breed in 2017.

DESIGNERCAT

LUXUS

ACF	ACFA	CAA	CCA-AFC	CCC	CCCA	CFA	FB	FIFe	GCCF	LOOF
							✓	✓		

NZCF	SACC	TICA	WCF
✓		✓	✓

Selection of the most important cat associations in the world: Australia, Canada, China, France, Germany, New Zealand, South Africa, UK and USA

Dwarf Cat

ATELIER KAYMAK®

Science PAPERS

Market price:
750.00 USD
–2,500.00 USD
(2023)

Recognized since 2009

LIFE EXPECTATION
10-15
YEARS

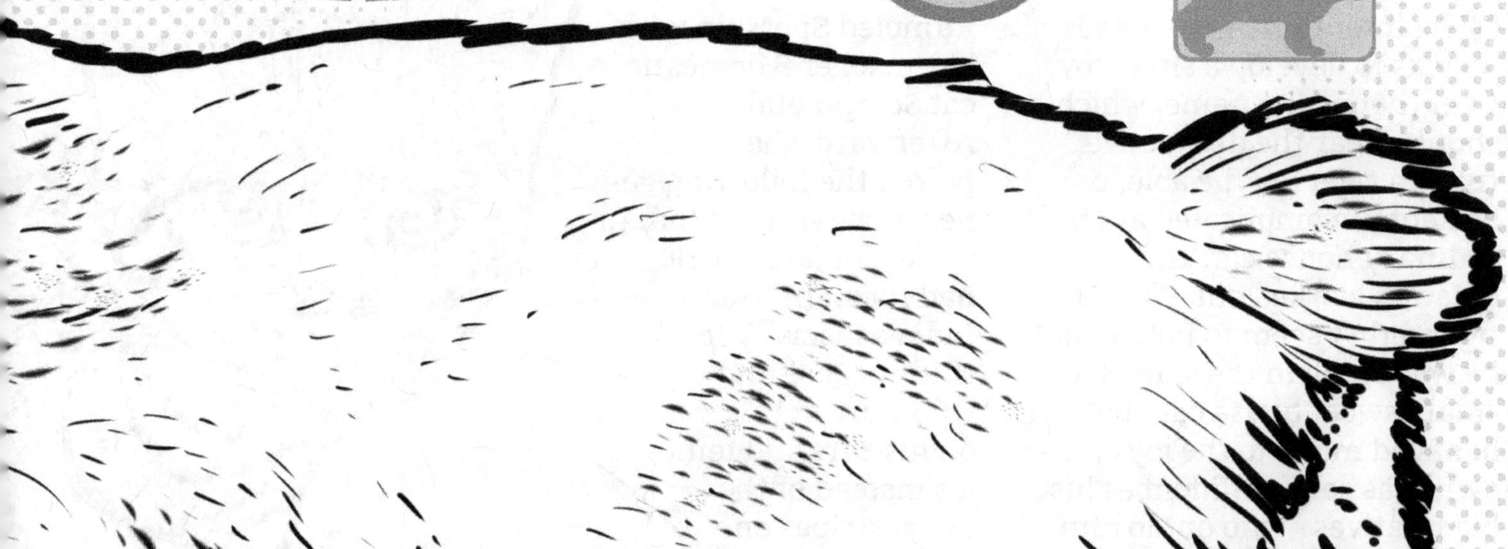

SHORT HAIR

Breeding start
1980s
United States

TOYGER

The intention of the breeder was to develop a small toy tiger (hence the name) which should wear the distinctive fur of a tiger and be able to be kept uncomplicated as an ordinary house cat. The calculation worked out: The mini tiger didn't seem to notice that he was stuck in the skin of a small, sweet house cat - but behaved much to the joy of his owner as self-confident as his big relatives - who outdo him many times over. The designed cat was bred in the 1980s by Judy Sudgen. Mrs. Sudgen followed in the footsteps of her mother, Jean Mill, who had previously bred the Bengal cat from a cross between a wild Asian Leopard cat and a domestic cat. Some years later, Mill's daughter crossed her spotted Bengal male Millwood Rumpled Spotskin with her mackerel domestic cat Scrapmetal. Afterward, she paired the following generation with the Indian street cat Jammu Blu, who had flown in from India especially for this purpose and who wore dots on his ears and temples instead of the usual stripes on these cats. Sudgen crossed cats with different stripe patterns to give their cats the look of their wild cousins from the jungle - which was absolutely unusual for cats until then. Although the fleshy, muscular Toyger cat has the appearance of a wild cat, it is a sociable cat with friendly nature.

ACF	ACFA	CAA	CCA-AFC	CCC	CCCA	CFA	FB	FIFe	GCCF	LOOF

NZCF	SACC	TICA	WCF
		✓	

Selection of the most important cat associations in the world: Australia, Canada, China, France, Germany, New Zealand, South Africa, UK and USA

LIFE EXPECTATION
10-15
YEARS

ATELIER KAYMAK

Science
PAPERS

283

Market price:
600.00 USD–50,000.00 USD
(2025)

TURKISH ANGORA

The Turkish Angora cat, formerly known as the ballerina of cats, is one of the oldest cat breeds. Louis XVI and his wife Marie-Antoinette noticed at that time every one of high rank should own at least one of these cats. According to genetic investigations, the graceful Turkish Angora cat is the mother of all Longhair cats in the world. She has her origin in Anatolia and in parts of the Asian-European Near East. It takes its name from the Turkish capital Ankara, which was formerly called Angora, and is located in the province of the same name - where this cat is still called Ankara kedisi today. Turkey recently appointed the Turkish Angora cat as the national cat. Turkish Angora cats are said to have been brought to Europe by merchants and returning crusaders as early as the 13th century. Turkish sultans began to gift Angora cats to the European nobility in the 1500s. Until the foundation time of the cat clubs in the 19th century, the Angora cat was very popular. The growing popularity of the Persian cat then caused confusion because no one really knew exactly how the Persian cat differed from the Angora cat. The Angora cat only reappeared after the Turkish state started a breeding program in its zoo in Ankara around 1900. The zoo was to take care of the preservation of the endangered Turkish Angora cat. However, because the stock of Angora cats remained relatively small, Angora cats were strictly protected and not or only very hesitantly given away. In 1963, however, an American couple managed to buy two Angora cats from the zoo. These formed the foundation of the breeding of Turkish Angora

Market price: 200.00 USD–3,500.00 USD (2025)

Breeding start 1960s United States

ACF	ACFA	CAA	CCA-AFC	CCC	CCCA	CFA	FB	FIFe	GCCF	LOOF
✓	✓	✓	✓			✓	✓	✓		

NZCF	SACC	TICA	WCF
✓		✓	

Selection of the most important cat associations in the world: Australia, Canada, China, France, Germany, New Zealand, South Africa, UK and USA

Weight 6-18 lbs

Recognized since 1968

cats in the USA. In 1973 the Turkish Angora cat was recognized by the CFA. Although only Angora cats of breeders from the USA and parts of Europe are recognized as race cats by the world-wide cat federations - the Turks regard all white lane cats on Turkish streets and bazaars as genuine Angora cats. Some Turks even reject the bred cat as too slim. The Turkish Angora cat is still one of the most popular breeds in the world of cat owners.

LIFE EXPECTATION
12-18 YEARS

(with pure white coat) Congenital hearing loss, deaf-

SHORT HAIR

BIG CAT

TURKISH VAN

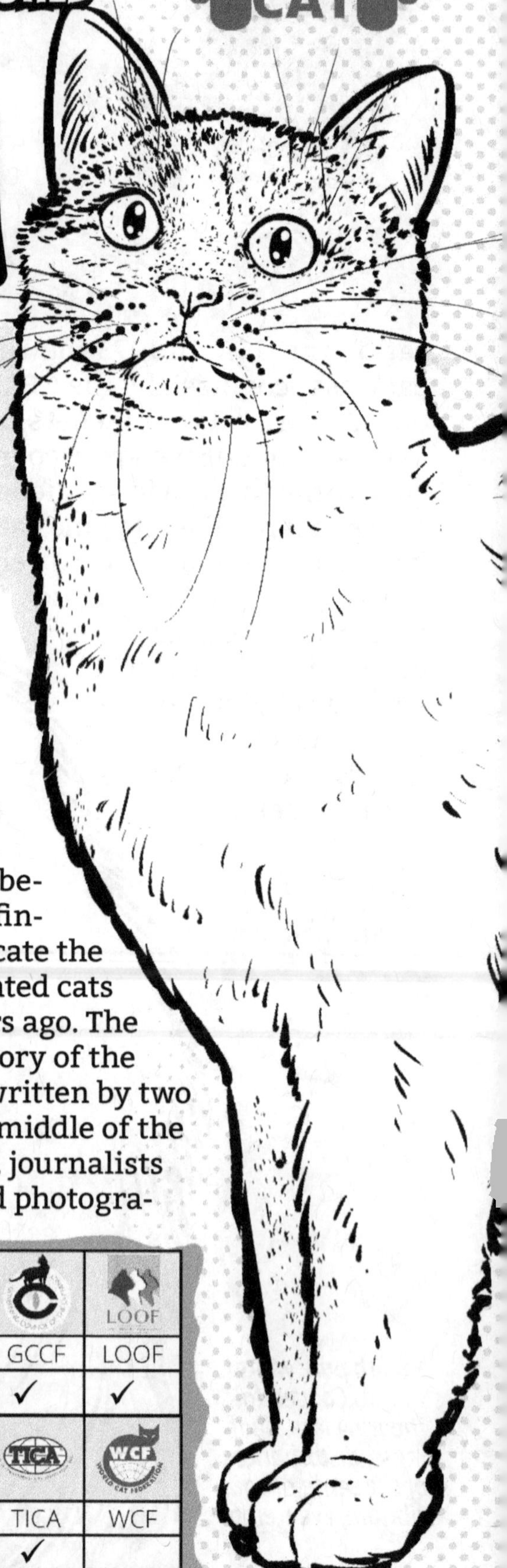

The Turkish Van cat was named after the East Anatolian province of the same name and the provincial capital Van - which owes their names to the nearby Lake Van. The lake is within sight of the biblical mountain Ararat, on whose summit Moses is said to have landed with his Noah's Ark. Lake Van is not only the largest lake in Turkey - the lake has several islands - it is also the largest soda lake in the world. According to the vernacular, after the water of the flood receded, cats climbed out of Noah's Ark and climbed down the mountain slopes to jump into Lake Van. According to another legend, Allah lovingly blessed the animals coming out of the ark with his hand-painted over their heads - which is why the animals had two reddish-brown markings growing on this spot. The Turkish Van cat has a long history because archaeological findings in the region locate the presence of domesticated cats in the time 7,000 years ago. The modern breeding history of the Turkish Van cat was written by two British women in the middle of the 20th century. In 1954, journalists Laura Lushingdon and photogra-

ACF	ACFA	CAA	CCA-AFC	CCC	CCCA	CFA	FB	FIFe	GCCF	LOOF
✓	✓	✓			✓	✓			✓	✓

NZCF	SACC	TICA	WCF
	✓	✓	

Selection of the most important cat associations in the world: Australia, Canada, China, France, Germany, New Zealand, South Africa, UK and USA

pher Sonia Halliday traveled to southern Turkey. At that time, the two women were employed by the Turkish Tourist Office and started their holidays. They reached the southernmost point of the country, the province of Hatay, which borders Syria. The two foreign women were treated with great hospitality and provided with safe accommodation everywhere. During a break, their host gave them two adult cats that the women liked: a local cat and a tomcat originally from Istanbul. Both cats wore white fur and had two reddish-brown markings on their heads. The two women set off on their journey with the two cats in their luggage. When they took a break at a river, the two cats jumped into the water to cool down and swam around. So, the Turkish Van cat came to her second name: swimming cat. After some attempts at recognition and naming (the Turkish Van cat is often confused with the Turkish Angora cat), the GCCF recognized the breed in 1969.

Market price:
150.00 USD–
1,500.00 USD
(2023)

Breeding start
1950s
United Kingdom

Recognized since
1969

(with pure white

LIFE EXPECTATION

TURKISH VANKEDISI ALIAS TURKISH VAN WHITE

The Turkish Vankedisi cat is essentially a lime white version of the Turkish Van cat. All other colors are categorized as Turkish Van. The bright white coat is the common standard for all cat breeds and is called the Van pattern. But the Turkish Vankedisi cat has one more special feature in stock: it can have different colored eyes. All Turkish Van cats, including the bright white Turkish Vankedisi cat, have a rather large stature. The cat can weigh up to 9 kg and can be 1.20 meters long from the tip of the nose to the end of the tail. The Van cat has a silky, water-repellent coat without an undercoat: a thin and airy summer coat and a dense and warm winter coat. Because very few Turkish Vankedisi cats still live in Turkey (in 1992, there were 92 cats in the whole country), state breeding programs try to maintain the population. The export of van cats from Turkey is therefore strictly prohibited and is punished with penalties

Breeding start
1990s
United Kingdom

ACF	ACFA	CAA	CCA-AFC	CCC	CCCA	CFA	FB	FIFe	GCCF	LOOF
									✓	

NZCF	SACC	TICA	WCF
		✓	✓

Selection of the most important cat associations in the world: Australia, Canada, China, France, Germany, New Zealand, South Africa, UK and USA

of up to US$ 40,000. The WCF granted the white Variety its own Category in 2009, but many associations do not recognize it as an independent breed.

Recognized since 2006

(with pure white coat) Congenital hearing loss, deafness, disturbances of the sense of equilibrium, Eye tremor

BIG CAT

Market price:
150,00 USD

LIFE EXPECTATION

SHORT HAIR

Breeding start
2000s
Ukraine

UKRAINIAN LEVKOY

The Ukrainian Levkoy cat is a new, unusual cat breed - which was developed only in 2000 by Elena Biriukova in Ukraine. Mrs. Biriukova then crossed a Donskoy cat with a Scottish fold cat - which manifested itself in the Offspring to a naked cat with folded ears. The Ukrainian Levkoy cat has a long, elegant body, narrow limbs, and a unique, doglike face with a prominent forehead and short muzzle. The resemblance to the dog is due to the angular dimensions of its head and the nasal bone, which merges almost seamlessly into the forehead. The Ukrainian Levkoy cat is regarded as a friendly, curious, and calm nature. The cat breed is recognized so far only by Ukrainian and Russian cat federations. Levkoy is the name of an herbaceous plant which is also called Mattiola or night violet. The violet-smelling plant is a member of the cruciferous family and has curved leaves that are similar in shape to the folding ears of the Ukrainian Levkoy cat. So far, the Ukrainian Levkoy cat has only been recognized in Ukraine (2005) and Russia (2010). In countries like Germany, the Animal Protection Act prohibits the breeding of the Ukrainian Levkoy cat.

HAIRLESS CAT

ACF	ACFA	CAA	CCA-AFC	CCC	CCCA	CFA	FB	FIFe	GCCF	LOOF

NZCF	SACC	TICA	WCF

Selection of the most important cat associations in the world: Australia, Canada, China, France, Germany, New Zealand, South Africa, UK and USA

ATELIER KAYMAK
Science
PAPERS
291
ORIGIN:
ASIA, UKRAINE
Recognized since 2005
Market price:
400.00 USD–
800.00 USD
(2023)
LIFE EXPECTATION
9-15
YEARS

ATELIER KAYMAK'S CREATIVE- & STUDY BOOKS
ALL CAT BREEDS OF THIS WORLD
by Nuesret Kaymak

SHORT HAIR

UKRAINIAN LEVKOY STRAIGHT

HAIRLESS CAT

Like other folding ear cats carrying the gene of the Scottish fold cat, some of the Ukrainian Levkoy cats develop wavy ears after a few weeks - while the ears of some other kittens just grow out. This is common in cat breeding because breeders are careful to maintain the cat population with a 50% share of normal cats. If only cats with a gene defect were bred - i.e., two cats with homozygous ear fold genetics (i.e., with two copies of the dominant ear fold gene), the cats would be born with hereditary diseases and associated health problems. In the case of the Ukrainian Levkoy straight cat - the Levkoy cat without a folding ear mutation - it is difficult for the viewer to believe that these Levkoy cats are different from their siblings with the folding ears. As with all naked cats, the Ukrainian Levkoy straight cat should be taken with caution when bathing. It can quickly happen that the oils produced by the skin are quickly washed away, and the cat's immune system is weakened. So far, the Ukrainian Levkoy straight cat has been recognized only in Ukraine (2005) and Russia (2010). In countries like Germany, the Animal Protection Act prohibits the the breeding of the Ukrainian Levkoy cat.

Market price:
400.00 USD–800.00 USD (2023)

Breeding start
2000s
Ukraine

ACF	ACFA	CAA	CCA-AFC	CCC	CCCA	CFA	FB	FIFe	GCCF	LOOF

NZCF	SACC	TICA	WCF

Selection of the most important cat associations in the world: Australia, Canada, China, France, Germany, New Zealand, South Africa, UK and USA

LIFE EXPECTATION
9-15 YEARS

ATELIER KAYMAK
Science
PAPERS
293
ORIGIN:
ASIA, UKRAINE
Recognized since 2005

SHORT HAIR

Breeding start
1980s
Soviet Union

URAL REX
SHORTHAIR

Recognized since 2006

LIFE EXPECTATION
12-15
YEARS

LUXUS

REX CAT

The Ural Rex cat is considered to be a Russian noble cat and is one of the rarest cat breeds in the world. In Russia, the locals of Sverdlovsk in the Ural Mountains claim that there were short-haired street cats already in the 1940s. But they were regarded as curiosities because of their appearance - and were not so popular. It is guaranteed that the official story about the origin of the Ural Rex cat began in 1994 in Ekaterinburg - and, as has so often been the case in the history of human civilization - about inbreeding. In the Poduhrovskih family house lived a black and white spotted domestic cat called Mura. Sometime in 1988, Mura became pregnant and gave birth to healthy kittens. Some kittens immediately attracted attention because they had an unusually dense and double curled coat. The Poduhrovskihs gave away all kittens except the little sweet tom cat Vasili, which they kept. When Vasili was ready, he was mated to a female cat because the Poduhrovskihs wanted more of these beautiful cats. But in the following litter, there were only kittens with normal, smooth fur. After that, they mated Vasili with his mother, Mura, and made Vasili the progenitor of all following Ural Rex cats. The Poduhrovskihs got their longed-for curly cats again. Mura gave birth to kittens with curly fur. It was clear that the Rex Gene is recessively inherited.

ACF	ACFA	CAA	CCA-AFC	CCC	CCCA	CFA	FB	FIFe	GCCF	LOOF

NZCF	SACC	TICA	WCF
			✓

Selection of the most important cat associations in the world: Australia, Canada, China, France, Germany, New Zealand, South Africa, UK and USA

Weight
8-12
lbs

Market price:
600.00 USD–
1,500.00 USD
(2023)

In the 90s, the Rex cats of the Poduhrovskihs were presented to the interested public at Russian cat exhibitions. In 2006, Elena Fedorenko presented the Ural Rex cat to the WCF Committee, which immediately recognized the cat as a new cat breed. In experimental breeding with other Rex varieties, it was noticed that the Ural Rex mutation was different from the other Rex mutations.

URAL REX LONGHAIR

The Ural Rex Longhair cat - or more precisely semi-long-hair cat - is one of the rarest cat breeds in the world - there are allegedly only about 65 specimens of it in the whole world. She is an affectionate and friendly cat who likes to live in the company of people. She has a strong body - but is still slim and very agile. She has a soft, silky, shiny coat, which is covered with loosely arranged, elastic double waves. Due to the S-shaped waves, the coat is so undemanding that only a moistened hand and a simple brush are sufficient for grooming. In comparison to the wedge-shaped head of the Cornish Rex cat - the Ural Rex Longhair cat has a round head and looks a little less elegant than this. Its almond-shaped eyes radiate friendliness and vitality. The WCF recognized the breed in 2006.

Breeding start 1980s Soviet Union

LIFE EXPECTATION 12-15 YEARS

REX CAT

LUXUS

ACF	ACFA	CAA	CCA-AFC	CCC	CCCA	CFA	FB	FIFe	GCCF	LOOF

NZCF	SACC	TICA	WCF
			✓

Selection of the most important cat associations in the world: Australia, Canada, China, France, Germany, New Zealand, South Africa, UK and USA

ATELIER KAYMAK
Science
PAPERS
297
ORIGIN:
ASIA, SOVIET UNION
Market price:
600.00 USD–
1,500.00 USD
(2023)
Recognized
since
2006

YORK CHOCOLATE

The York Chocolate line began in 1983 in a goat dairy, where a promising mating of Janet Cheifari's black-and-white spotted, curly farmer cat Blacky - and the neighborhood living, fat tomcat Smokey took place. The whole thing must have happened secretly because the dairy staff was amazed when two months later, they found a fluffy litter of cats in a hidden corner under a milking stool. One of the kittens was strikingly different: it was a sugar-sweet, chocolate-brown kitten - which all dairy employees fell in love with and called Brownie. Brownie immediately attracted the attention of the owner of the dairy, Janet Chiefari - who also fell in love with the kitten. Moreover, Mrs. Chiefari saw herself obliged to breed this cat. Brownie looked really lovely and had charm. So, it turned out quite fast that she was already producing her own litter of kittens next summer. Also, this time, one of the kittens differed from the others: she wore a black coat of long black hair and a deep brown undercoat. Minky was mated to her mother a year later, leading to the birth of Teddy Bear, a brown male, and Cocoa - a brown-white female cat. The breeding was continued, and many sweet cats were born. In 1987 Mrs. Chiefari had 27 cats, not counting the many cats that had been poached from her by visitors and friends.

Breeding start
1990s
United States

LIFE EXPECTATION
13-15
YEARS

Recognized since
1990

ACF	ACFA	CAA	CCA-AFC	CCC	CCCA	CFA	FB	FIFe	GCCF	LOOF
										✓

NZCF	SACC	TICA	WCF
			✓

Selection of the most important cat associations in the world: Australia, Canada, China, France, Germany, New Zealand, South Africa, UK and USA

ATELIER KAYMAK

Science PAPERS

Market price:
600.00 USD–
1,200.00 USD
(2023)

IMPRINT

+++ Don't Forget to Leave a Review on the Web +++

Author

Nuesret Kaymak is a Sagittarius-Ascendant-Sagittarius and works as a creative for PR, advertising and TV. Besides, he writes and illustrates books about animals, nature and history, his hobbyhorse.

Legal matters

Fonts used in this book

- Caecilia
- Open Sans
- Ghoulish Fright

Sources

- acf.asn.au
- acfacat.com
- alleycat.org
- American Association of Feline Practitioners (AAFP)
- American cat Fanciers Association (ACFA)
- animalwised.com
- Australian cat Federation (ACF)
- Cambridge University
- Canadian cat Association/Association Féline Canadienne (CCA-AFC)
- Cat Aficionado Association (CAA)
- Cat Fanciers' Association (CFA)
- catvets.com
- cca-afc.com
- cccofa.com.au
- cfa.org
- Chat Canada cats (CCC)
- chatscanadacats.ca
- Columbia University
- Co-Ordinating Cat Council of Australia (CCCA)
- Cornell University
- en.wikipedia.org
- Fédération Internationale Féline (FIFe)
- Fédération pour la gestion du Livre Officiel des Origines Félines (LOOF)
- Felis Britannica (FB)
- felisbritannica.com
- fifeweb.org
- gccfcats.org
- Governing Council of the cat Fancy (GCCF)
- icatcare.org
- International Society of Feline Medicine (ISFM)
- loof.asso.fr
- National Center for Biotechnology Information, U.S. National Library of Medicin
- natureasia.com
- New Zealand cat Fancy (NZCF)
- nzcf.com
- Smithsonian
- Southern Africa cat Council (SACC)
- The International cat Association (TICA)
- tica.org
- tsacc.org.za
- University of California
- University of Michigan
- University of Washington
- wcf-online.com
- wcf-online.de
- World cat Congress (WCC)
- World cat Federation (WCF)
- worldcatcongress.org

Bibliography

The Complete cat Book. Gebhardt, Richard H. Howell Book House. New York, 1991.

The Royal Canin Encyclopedia. Groupe Royal Canin. Paris, 2000.

Cat World, a Feline Encyclopedia, Morris, Desmond. Penguin Random House, Ebury. London, 1996.

The Cornell Book of cats. Siegal, Mordecai and Cornell University. Random House, Ballantine Books, Villard Books. London, 1989.

Die domestizierte Katze. Turner, Dennis, Bateson, Patrick. Albert Müller Verlag. Zürich, 1988.

The Encyclopedia of the cat. Fogle, Bruce. Dorling Kindersley. London, 1997.

Legacy of the cat. Stephens, Gloria. Chronicle Books. San Francisco, 1990.

Outline of cat Anatomy. Gilbert, Stephen G. University of Washington Press. Seattle, 2000.

The Complete cat Encyclopedia. Pond, Grace. Crown Publishers. New York, 1972

Rassekatzen: die 100 Katzenrassen. Morris, Desmond. Heyne Verlag. München, 1999.

Cat Breeds of the World. Morris, Desmond. Penguin, Viking. London, 1999.

Our cats and all about them. Weir, Harrison. Fanciers' Gazette. London, 1892.

The Cat Book, cats of Historical Distinction. Walker-Meikle, Kathleen. Bloomsbury Publishing, Old House Books. London, 2015.

Beastly possessions: animals in Victorian consumer culture. Amato, Sarah. University of Toronto Press. Toronto, 2015.

Animals and society: an introduction to human-animal studies. Columbia University Press. New York, 2012.

Everybody's cat book. Champion, Dorothy Bevill. Lent & Graff Company. New York, 1909.

www.ingramcontent.com/pod-product-compliance
Lightning Source LLC
LaVergne TN
LVHW060445061126
851857LV00016B/1737
* 9 7 8 3 9 6 1 8 3 0 3 1 2 *